New Approaches to Reliability Qualification
of Semiconductor Components
under Varying and Progressive Stresses

UNIVERSITÄT DER BUNDESWEHR MÜNCHEN
Fakultät für Elektrotechnik und Informationstechnik

New Approaches to Reliability Qualification of Semiconductor Components under Varying and Progressive Stresses

Alexander Hirler

Vollständiger Abdruck der von der Fakultät für
Elektrotechnik und Informationstechnik
der Universität der Bundeswehr München
zur Erlangung des akademischen Grades eines

Doktor-Ingenieurs

genehmigten Dissertation.

Gutachter: 1. Univ.-Prof. Dr.-Ing. Walter Hansch

2. Univ.-Prof. Dr.-Ing. Rainer Marquardt

Die Dissertation wurde am 14.01.2021 bei der
Universität der Bundeswehr München eingereicht und durch die Fakultät für
Elektrotechnik und Informationstechnik am 09.06.2021 angenommen.
Die mündliche Prüfung fand am 28.06.2021 statt.

Bibliografische Information der Deutschen Nationalbibliothek
Die Deutsche Nationalbibliothek verzeichnet diese Publikation in der Deutschen Nationalbibliografie; detaillierte bibliographische Daten sind im Internet über http://dnb.d-nb.de abrufbar.
1. Aufl. - Göttingen: Cuvillier, 2021
Zugl.: Universität der Bundeswehr München, Diss., 2021

Nonnenstieg 8, 37075 Göttingen
Telefon: 0551-54724-0
Telefax: 0551-54724-21
www.cuvillier.de

1. Auflage, 2021
Gedruckt auf umweltfreundlichem, säurefreiem Papier aus nachhaltiger Forstwirtschaft.

ISBN 978-3-7369-7520-0
eISBN 978-3-7369-6520-1

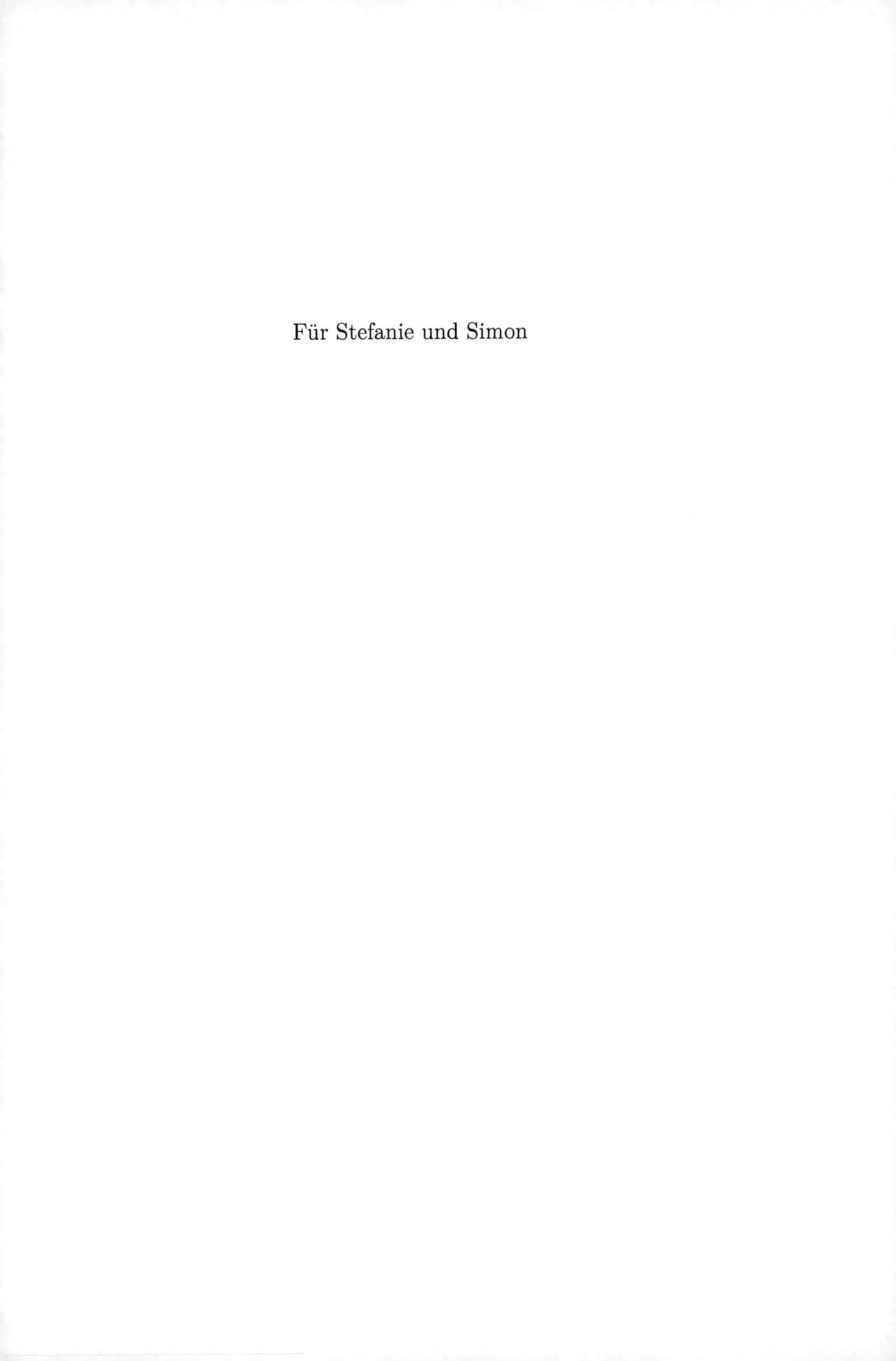

Für Stefanie und Simon

Abstract

In the present work, urgent issues in the reliability qualification of semiconductor devices are addressed, which particularly affect value chains such as those in the automotive industry. These have particularly high requirements for long lifetime and low failure rates of their products, which are additionally exposed to more extreme operating and environmental conditions than in most other areas of application. In particular, the question arises on how to assess a product or semiconductor technology against the requirement of an application-specific mission profile with multiple non-constant stressors.

For this purpose, the behavior of failure distributions under varying and progressive stress loads is investigated and described using cumulative damage models. For the first time, the industry-wide approach of transforming non-constant mission profiles into effective constant stress and test conditions for reliability assessment and qualification can be physically justified and substantiated with measurement data. This stress transformation is exemplified using the time-dependent dielectric breakdown (TDDB) failure mechanism as an example, and then extended to include the use of multi-dimensional mission profiles and interdependent stressors.

First, the fundamentals of statistics, data analysis, and reliability methodology necessary for reliability studies are presented and discussed. These are then specified for the semiconductor failure mechanism TDDB and applied practically, using the example of an industry-oriented technology qualification, and in direct comparison with reliability data from the semiconductor manufacturer GLOBALFOUNDRIES.

In order to be able to transfer mission profiles and stress histograms into effective stress conditions, the properties of three well-known and accepted cumulative damage models, namely the cumulative exposure (CE) model, the tampered random variable (TRV) model and the tampered failure rate (TFR) model, are examined in this regard. The detailed analysis of damage accumulation for alternating step-stress reveals that equivalent effective failure distributions emerge only when CE and TRV models are applicable. The derivation of the effective acceleration factor shows that it can be calculated as weighted harmonic mean of the acceleration factors of the individual stress steps. In general, the TFR model does not allow these conclusions.

Using alternating step-stress measurements for the first time for the TDDB failure mechanism, the CE and TRV models are confirmed for the two stressors voltage and temperature, as well as the equivalence of cyclical step-stress with an effective constant stress. These measurements are performed on university metal–oxide–semiconductor (MOS) capacitors as well as on GLOBALFOUNDRIES 22FDX® state-of-the-art transistors.

The applicability of deriving effective stresses for multi-dimensional mission profiles of linked and interdependent stressors is demonstrated using voltage and temperature stress for TDDB as an example. The examination of a shown application example from the automotive industry allows the conclusion that neglecting linked stressors has a significant impact on determined lifetimes and the respective required test times. In this study, the extent of the discrepancy between the determined times is at least an order of magnitude.

Finally, the results obtained in the field of cumulative damage are also applied to ramp-stress. The four most accepted TDDB acceleration models are analyzed in detail for the transformation of failure distributions from constant stress to ramp-stress and vice versa. These models are: 'exponential E model', 'exponential $\sqrt{E}$ model', 'power-law U model' and 'exponential $1/E$ model'. It is demonstrated that the obtained findings can be applied to use reliability studies with ramp-stress for model verification and to determine model parameters with less time and experimental effort.

Zusammenfassung

In der vorliegenden Arbeit wird auf dringende Problemstellungen in der Zuverlässigkeitsqualifizierung von Halbleiterbauelementen eingegangen, die insbesondere Wertschöpfungsketten wie die der Automobilindustrie betreffen. Diese haben besonders hohe Anforderungen an lange Lebenszeiten und geringe Ausfallraten ihrer Produkte, die zusätzlich extremeren Betriebs- und Umweltbedingungen ausgesetzt sind, als in den meisten anderen Anwendungsbereichen üblich. Im Speziellen geht es um die Fragestellung, wie ein Produkt oder eine Halbleitertechnologie gegen die Anforderung eines applikationsspezifischen Einsatzprofils mit mehreren nichtkonstanten Stressoren abgesichert werden kann.

Dazu wird das Verhalten von Ausfallverteilungen unter nichtkonstanten und progressiven Stressbelastungen mithilfe von Modellen der Schadensakkumulation untersucht und beschrieben. Erstmals kann das industrieweit verwendete Vorgehen, bei dem nichtkonstante Einsatzprofile in effektive konstante Stress- und Testbedingungen transformiert werden, bei der Auslegung von Zuverlässigkeit und der Qualifizierung physikalisch begründet und mit Messdaten belegt werden. Diese Stresstransformation wird am Beispiel des Fehlermechanismus 'zeitabhängiger dielektrischer Durchbruch (TDDB)' exemplarisch gezeigt und anschließend auch auf den Einsatz mehrdimensionaler Einsatzprofile und voneinander abhängiger Stressoren ausgebaut.

Zunächst werden die für Zuverlässigkeitsuntersuchungen notwendigen Grundlagen der Statistik, Datenauswertung und Zuverlässigkeitsmethodik vorgestellt und diskutiert. Diese werden daraufhin für den Halbleiterfehlermechanismus TDDB konkretisiert und am Beispiel einer industrienahen Technologiequalifizierung praktisch und im direkten Vergleich mit Zuverlässigkeitsdaten des Halbleiterherstellers GLOBALFOUNDRIES angewandt.

Um Einsatzprofile und Stresshistogramme in effektive Stressbedingungen überführen zu können, werden diesbezüglich die Eigenschaften der drei bekanntesten Modelle für Schadensakkumulation, nämlich das 'Cumulative Exposure (CE) Modell', das 'Tampered Random Variable (TRV) Modell' und das 'Tampered Failure Rate (TFR) Modell', untersucht. Bei der theoretischen Betrachtung der Schadensakkumulation von alternierendem Stufenstress wird gezeigt, dass sich nur bei einer zulässigen Anwendbarkeit des CE und TRV Modells äquivalente effektive Ausfallverteilungen herausbilden. Die Herleitung des effektiven Beschleunigungsfaktors zeigt, dass sich dieser als gewichtetes harmonisches Mittel der Beschleunigungsfaktoren der einzelnen Stresstufen berechnen lässt. Das TFR Modell lässt diese Schlussfolgerungen im Allgemeinen nicht zu.

Anhand erstmalig durchgeführter alternierender Stufenstressmessungen für den Fehlermechanismus TDDB wird das CE und TRV Modell für die beiden Stressoren Spannung und Temperatur sowie die Gleichwertigkeit von zyklischem Stufenstress mit einem effektiven konstanten Stress bestätigt. Diese Messungen erfolgen an eigens hergestellten Metall–Oxid–Halbleiter (MOS) Kondensatoren sowie an Transistoren der GLOBALFOUNDRIES 22FDX® Technologie, die dem aktuellsten Stand der Technik entsprechen.

Die Anwendbarkeit der hergeleiteten Effektivierung der Stufenbelastungen für mehrdimensionale Einsatzprofile aus verknüpften und voneinander abhängigen Stressoren wird am Beispiel von Spannungs- und Temperaturstress für TDDB nachgewiesen. Die Betrachtung eines gezeigten Anwendungsbeispiels aus der Automobilindustrie lässt die Schlussfolgerung zu, dass eine Nichtbeachtung von verknüpften Stressoren erhebliche Auswirkungen auf ermittelte Lebenszeiten und die dafür notwendigen Testzeiten hat. In dieser Untersuchung beträgt das Ausmaß der Diskrepanz zwischen den ermittelten Zeiträumen mindestens eine Größenordnung.

Abschließend werden die gewonnen Ergebnisse auf dem Gebiet der Schadensakkumulation auch auf Rampenbelastungen übertragen. Die vier anerkanntesten Beschleunigungsmodelle werden für die Transformation von Ausfallverteilungen unter konstantem und Rampenstress einer ausführlichen Analyse unterzogen. Diese Modelle sind: Das 'exponentielle E Modell', das 'exponentielle $\sqrt{E}$ Modell', das 'Potenzgesetz U Modell' und das 'exponentielle $1/E$ Modell'. Es wird gezeigt, wie sich die gewonnen Erkenntnisse unter anderem dafür einsetzen lassen, Zuverlässigkeitsuntersuchungen mit Rampenstress für die Modellüberprüfung zu nutzen und Modellparameter mit geringerem zeitlichen und experimentellen Aufwand zu ermitteln.

Contents

1 Introduction

Just as mechanical constructs have lifetime and reliability limits, electronic components in general, and semiconductor devices in particular, can fail when reaching their technical limits as well. This can lead to a malfunctioning or complete failure of the superior system, may it be a consumer product, a leading-edge industrial application, or a safety-critical system in cars and airplanes. In order to test the coverage of customer's requirements by the reliability limits of the device, reliability qualification is performed.

The current situation in reliability qualification and testing of semiconductor devices is challenging, because already well-established qualification test-plans are confronted with novel and more demanding lifetime requirements than ever before. Especially in the automotive sector, with its mega-trends of driver assistance systems, autonomous driving, electric mobility and car connectivity on their way into the market, new lifetime and failure rate requirements appear that are new to the automotive supply chain. Thus, qualification plans that have fixed test conditions and test times cannot withstand the current evolution. In conclusion, to satisfy the needs for adequate reliability qualification, new approaches are necessary and inevitable for assuring failure free products in the future.

> “*If an automotive [original equipment manufacturer (OEM)] goes to their tier 1 or 2 and asked for a system with a failures in time (FIT) rate of 1 per billion hours of operation, and you take an existing 100 million gate chip and place the same requirement on that supplier – we don't even really know what that means. How do I take a 100 million gate chip and determine that is has a FIT of 1 per billion hours of operation? When I first found out how this is being done today, I tried to turn off all of the electronics in my car. If that was how they determined that it was safe, then I was better off without the electronics.*”
>
> — Apurva Kalia, vice president of Research & Development in the System and Verification group of Cadence. [Bai18]

A step has to be made towards mission profile aware reliability and robustness validation, which is a thorough evaluation of all reliability aspects of the technology and product as well as the specific requirements imposed on it. This includes an advanced understanding and modeling of the real environmental and functional stresses on the product, a concept of time-saving and more elaborated testing, suitable reliability theory and models, and of course empirical validation of all the aforementioned.

Chapter 2 provides the necessary foundations in the areas of statistical description of failure behavior and reliability predictions based on accelerated testing in order to enable the assessment of the reliability investigations and statements appropriately. The introduced concepts are clearly illustrated and pitfalls in the implementation are pointed out.

Chapter 3 elaborates the presented methods on the basis of the failure mechanism of time-dependent dielectric breakdown. The current state of research on the failure physics and the methods used to investigate the reliability behavior of this mechanism are presented in detail. They are illustrated comprehensively on the basis of in-house measurements and in comparison with qualification data from a semiconductor manufacturer.

Chapter 4 outlines the reasons for the importance of processing and evaluating mission profile based reliability requirements for the automotive industry and the fact that fundamental aspects have not yet been fully described and empirically verified. First using well-known cumulative damage models, the failure behavior of semiconductor components under alternating step-stresses is analyzed and essential differences are identified. The gained insights are then used to develop a method to transform mission profile stresses into effective stress conditions. This is demonstrated and proven by cyclical stress measurements for the stressors voltage and temperature, which were conducted for the first time. Furthermore, findings are made about the applicability of cumulative damage models that have not previously been reported in this context. Finally, the coupling of stressors in multi-dimensional mission profiles is examined in detail and the consequences of consideration and neglect of these interdependencies are clearly illustrated using a real-world example.

Chapter 5 highlights the specific challenges that can be addressed using more sophisticated reliability testing methods. In particular, voltage ramp-stress testing can provide significant benefits for wafer level reliability in technology development and production. The individual stress transformation characteristics of the four voltage acceleration models used for dielectric breakdown are analyzed comprehensively and the implications for the scale and shape parameters of the ramped failure distribution are elaborated. Finally, the capabilities of ramp-stress measurements for model verification and parameter fitting are presented and compared to the standard methods currently used in reliability qualification.

Chapter 6 concisely summarizes the results and contributions of this work. Potential fields for further research and new as-yet unsolved issues are identified and outlined.

2 Theory of Reliability Testing

To characterize a given component in terms of reliability prediction, an understanding of statistical distributions and reliability methodology is necessary. Hence, this chapter will give an introduction into the theory of reliability testing. Especially, the statistical methods and distributions used in this work will be described in detail.

For a greater insight into the topic of reliability, the works of A. Strong [Str09] and J. McPherson [McP19] are recommended as supplementary literature. As an excursion into statistical methods, the respective handbook of the American National Institute of Standards and Technology (NIST) [NIS13] is endorsed.

2.1 Statistical Description of Failures

To begin with, fundamental terms, definitions, and concepts are introduced to facilitate the understanding of the following chapters.

2.1.1 Basic Concepts of Failure Statistics

When expressing failure behavior in a mathematical way, the first encountered term is the failure function $F(t)$. It simply describes the probability that a device will fail at or before time t. It is expressed as cumulated failed percentage of a given population and is generally a continuously increasing function. Contrarily, the survival or reliability function $R(t)$ states the probability that a device has not failed and thus is still working until time t. Failure and survival are considered to be complementary by nature:

$$F(t) = 1 - R(t) \tag{2.1}$$

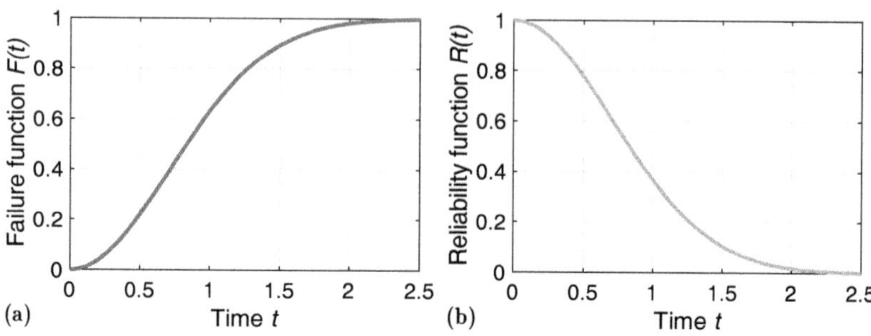

Figure 2.1: Complementary behavior of **(a)** the failure function and **(b)** reliability function for the same population.

Both functions are illustrated in Fig. 2.1 and exhibit function values ranging from 0 to 1 and vice versa, until the entire population has failed. [Str09]

Due to its cumulative behavior, the failure function $F(t)$ is also called cumulative distribution function (CDF). When explicitly talking about the empirically obtained failure function, the terms empirical CDF or empirical distribution function (EDF) are often used. As depicted in Fig. 2.2a, the EDF exhibits steps-like behavior due to the discrete nature of measured data. The CDF, whether derived theoretically or fitted from data, is usually displayed as a continuous function.

When failures are not described in a cumulative way but rather as failures in each period of time, the derived progression is called probablity density function (PDF) $f(t)$. It is the derivative of the CDF $F(t)$ and can be for example displayed as a histogram for empirical data:

$$f(t) = \frac{\mathrm{d}F(t)}{\mathrm{d}t} \qquad \text{or} \qquad F(t) = \int_0^t f(x)\,\mathrm{d}x \tag{2.2}$$

The PDF is usually the most encountered function when dealing with probability distributions in mathematical analysis because of its convenient properties as density function of the distribution (see Fig. 2.2b). [Str09]

Another important concept in statistics is that of the instantaneous failure rate (FR) or hazard function $h(t)$. It states the probability of the surviving specimens at time t to fail within the next time frame. It is expressed as the PDF $f(t)$ divided by the remaining survivors, which is given by the reliability function $R(t)$:

$$h(t) = \frac{f(t)}{1 - F(t)} = \frac{f(t)}{R(t)} \tag{2.3}$$

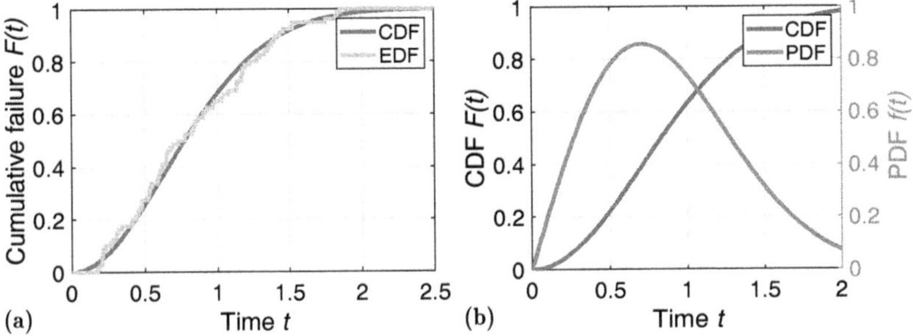

Figure 2.2: **(a)** CDF of a Weibull distribution and EDF of randomly generated data, often depicted as a stairstep graph. **(b)** CDF and its derivation PDF.

It is often used for describing the reliability behavior in certain time spans of a system's course of life or when fail probability of the current state and the immediate future is discussed. The most prominent application of the hazard function, the so-called *bathtub curve*, is described in the following section 2.1.2. [Str09]

For certain time frames, the average failure rate (AFR) $\langle h(t) \rangle$ can be an interesting figure of merit when comparing products from different suppliers for fails during equal time intervals, like the first year or the required lifetime of the component or system. The AFR is defined as the total number of fails within a given interval of time. In the reliability community, AFR represents a rate and is usually given in units of failures in time (FIT), which is fails per billion device hours ($^{10^{-9}}/_{\mathrm{h}}$):

$$\langle h(t) \rangle = \frac{\int_0^t h(x)\,\mathrm{d}x}{\int_0^t \mathrm{d}x} = \frac{1}{t}\int_0^t \frac{f(x)}{1-F(x)}\,\mathrm{d}x = \frac{1}{t}\ln\left[\frac{1}{1-F(t)}\right] \tag{2.4}$$

Along with the AFR, there is also a reciprocal expression of this rate called mean time between failures (MTBF) (MTBF $= {}^{1}/_{\langle h(t) \rangle}$), which acts as a very important parameter for reliability statements as well.

The last concept in this introductory section will be that of the acceleration factor (AF). The dimensionless quantity AF is defined as the ratio between two failure times, often named time-to-failure (TTF), which are corresponding failure times of the same or different distributions:

$$\mathrm{AF} = \frac{\mathrm{TTF_{nop}}}{\mathrm{TTF_{acc}}} \tag{2.5}$$

In manner of speaking, the corresponding stress level of $\mathrm{TTF_{acc}}$ is said to be AF-times more accelerated than the stress level under normal operating conditions of

$\mathrm{TTF}_{\mathrm{nop}}$ (if AF > 1). In case of AF < 1, the relation of these stress levels would be referred to as deceleration — or the fraction would simply be inverted in order to conform to the prior statement. The application of AF is universal and can be used for reliability requirements, measurements, and extrapolations alike. Hence, there is no consistent definition covering all individual cases.

2.1.2 The Bathtub Curve

The most used image for visualizing the reliability behavior of any system or device is probably the so-called *bathtub curve* for reliability. It depicts the relation of instant failure rate $h(t)$ to lifetime and consists of three different segments. These are shown in Fig. 2.3a–c as falling, constant, and rising regime, which give it, because of the bathtub-like appearance, its descriptive name. Regardless of whether semiconductor devices, living beings, or structural components are being described by this curve, generally speaking, these main characteristics apply to all statistical failure mechanisms.

The first segment on the left is the *early fails* or *infant mortality* region, named early failure rate (EFR). It is caused by production and manufacturing related (gross) defects that results in early fails of the devices even under normal operating conditions. The failure rate, which is often given units of FIT, is characterized by

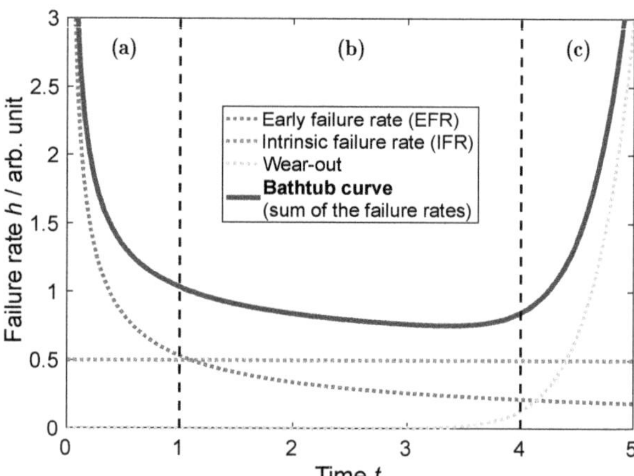

Figure 2.3: The *bathtub curve* has three distinctive regions (from left to right): **(a)** Infant mortality region with strongly decreasing EFR, **(b)** operating life with an overall low IFR, and **(c)** end-of-life when the wear-out failure rate increases until all specimens have failed.

a considerably high starting value and a following rapid decrease usually within the warranty period of the product.

In order to minimize the impact early fails have on customer satisfaction, one method for the manufacturer is to screen out such material and production defects by a so-called *burn-in* stress, where weak devices will fail under short and highly accelerated stress conditions. However, the burn-in conditions have to be precisely adjusted to not cause unnecessary degradation and premature aging of the remaining specimens. The other approach, more suitable for high-value products, is a specially tailored guarantee time to absorb the consequences of early fails after shipment to the customer. [Str09]

After the failure rate drop of the initial EFR regime, a stable and preferably low intrinsic failure rate (IFR) bottom segment of the bathtub curve emerges — this is considered the *operating life*. This area presents the maximal capabilities of the production and failures are due to intrinsic weaknesses, such as very small defects in the material [McP19]. While the characteristic of the IFR region is described as having a constant failure rate in general, the failure rate is usually slightly decreasing over the span of the operating life [Str09].

For obvious reasons, the operating life regime with the lowest failure rate should be extended as long as possible before the upcoming *wear-out* regime with a sharp increasing failure rate sets in. Even the best devices will start to fail at some point due to wear-out effects. This is not correlated to any manufacturing or preexisting material defects but rather the utmost achievable reliability of the used materials and design-rules under the experienced use conditions [McP19]. This region defines the *end-of-life* of the product as eventually all devices will fail. Therefore, the manufacturer strives to determine the onset of the wear-out accurately to ensure that the reliability requirements of the customer are met with a high degree of certainty.

In conclusion, understanding and subsequently controlling all three segments of the bathtub curve during product development is key to good product reliability.

2.1.3 Failure Distributions

When using the bathtub curve for reliability modeling, the curve's three regimes, i.e. *early fails*, *operating life*, and *wear-out*, feature different failure characteristics and can therefore be described by separate failure distributions. One distribution which is capable of modeling all regions of the bathtub curve is the renowned Weibull distribution.

Weibull Distribution

In 1951, Waloddi Weibull published a distribution function with wide applicability and illustrated his claim with several examples. This included fitting data from yield strength and fatigue life of steel, fiber strength of cotton, and size distributions of fly ash, beans, and humans. Weibull was aware that this function is not always valid, but hoped for it to be of good service in some cases. [Wei51]

The Weibull distribution in its two-parametric form with scale parameter α and shape parameter β is valid for times t and parameters α, β all > 0 and given by the following formulas:

$$\text{PDF}: \quad f(t) = \left(\frac{\beta}{\alpha}\right) \cdot \left(\frac{t}{\alpha}\right)^{\beta-1} \cdot \exp\left[-\left(\frac{t}{\alpha}\right)^{\beta}\right] \tag{2.6a}$$

$$\text{CDF}: \quad F(t) = \int_0^t f(x)\,\mathrm{d}x = 1 - \exp\left[-\left(\frac{t}{\alpha}\right)^{\beta}\right] \tag{2.6b}$$

$$\text{FR}: \quad h(t) = \frac{f(t)}{1-F(t)} = \left(\frac{\beta}{\alpha}\right) \cdot \left(\frac{t}{\alpha}\right)^{\beta-1} \tag{2.6c}$$

$$\text{AFR}: \quad \langle h(t) \rangle = \frac{1}{t} \ln\left[\frac{1}{1-F(t)}\right] = \left(\frac{1}{\alpha}\right) \cdot \left(\frac{t}{\alpha}\right)^{\beta-1} \tag{2.6d}$$

The scale parameter α is the pivot point of the distribution, as can be seen in Fig. 2.4a, and corresponds to the moment in time when $F(\alpha) = 1 - \exp\left[-\left(\alpha/\alpha\right)^{\beta}\right] = 1 - \mathrm{e}^{-1} = 0.63212056\ldots$ and thus approximately 63 % of the specimens have failed by that time. The scale parameter α is therefore also known as $\boldsymbol{t_{63}}$. Due to the fact that $\boldsymbol{t_{63}}$ is independent of the shape parameter β, it is called the *characteristic life* of the Weibull distribution and also determines the spread of the distribution as depicted in Fig. 2.4c. Since α is a Weibull distribution specific parameter, it is usually used in the context of reliability and distribution modeling, whereas the denotation $\boldsymbol{t_{63}}$ is generally used to describe experimental data which are considered to be Weibull distributed. [Str09, Nel82]

Depending on the shape parameter β, the Weibull distribution can change its characteristics and can match or closely resemble other distributions, as can be seen in Fig. 2.4b. For the case that $\beta = 1$, the Weibull distribution reduces to a simple exponential distribution, for $\beta = 2$ it equates to the Rayleigh distribution, and for larger values $\beta \gtrsim 10$ it approaches the smallest extreme value distribution. In the mid-range of $\beta \approx 3.6$ the Weibull distribution can resemble the normal distribution with the largest deviation at about 8 % cumulated fails [Nel82]. Hence the Weibull shape parameter β has the ability to produce left and right skewed distributions, which makes it quite convenient for fitting different reliability data of unknown nature as well as the different sections of the bathtub curve. As a rough guideline,

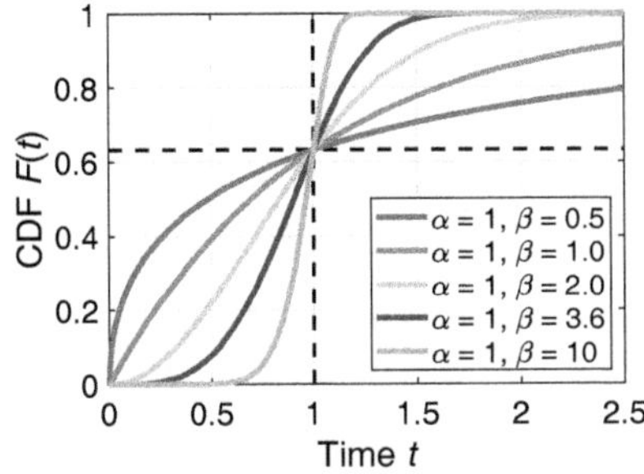

(a) CDF with different Weibull shape parameter

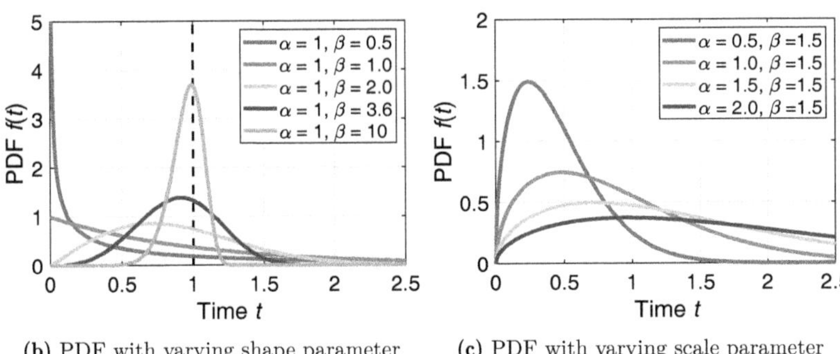

(b) PDF with varying shape parameter

(c) PDF with varying scale parameter

Figure 2.4: Different varying scale and shape parameters of the Weibull distribution. Vertical dashed lines indicate fixed t_{63} lifetimes.

values of $\beta \leq 0.8$ suggest extrinsic failure mechanisms which most likely cause early fails and $\beta \geq 2.4$ typically indicates intrinsic wear-out mechanisms [JED16].

Primarily, the application and original purpose of the Weibull distribution is to model *weakest-link* type failure mechanisms and problems. That is the probability that a system made of many components, which are subject to the same failure distribution, fails due to the failure of a single part or segments — like a chain fails due to the failure of one of its chain links. Examples for Weibull distributed events are static or dynamic strengths, electrical insulation breakdowns in dielectrics, and even the death of living systems. The weakest-link model also applies to the reliability of assembled systems which will fail if a single component fails and thus can be described by the Weibull distribution. [Wei51, McP19, Str09, Nel82]

Lognormal Distribution

The second important reliability distribution is the lognormal or logarithmic normal distribution. As the name indicates, it is derived from and in close relation to the normal distribution. If Y is a normal distributed random variable, then $Z = \mathrm{e}^Y$ is lognormal distributed with the scale or median parameter μ, and the shape parameter σ [Str09]. The parameter μ is also called *log mean* and is therefore the mean of the logarithm of time t. The mean of lifetime, when 50 % of the population have failed, which is usually denoted as t_{50}, is hence related to μ as $\mu = \ln(t_{50})$ (see Fig. 2.5a). Analogously, the shape parameter σ is called *log standard deviation* and is therefore the standard deviation of the logarithm of time t, which can be approximated by $\sigma \approx \ln(t_{50}) - \ln(t_{16})$. This of course implies that, unlike time t, the parameters μ and σ are not times but instead dimensionless numbers [Nel82]. Note that the formulas and descriptions of the lognormal distribution in this work use the natural logarithm ln( ) whereas other authors may use the common logarithm log( ).

The distribution functions of the lognormal distributions are defined with positive t and σ [McP19, Bro12]:

$$\text{PDF}: \qquad f(t) = \frac{1}{\sigma t \sqrt{2\pi}} \cdot \exp\left[-\left(\frac{\ln(t) - \mu}{\sigma\sqrt{2}}\right)^2\right] \tag{2.7a}$$

$$\text{CDF}: \qquad F(t) = \Phi\left(\frac{\ln(t) - \mu}{\sigma}\right) \tag{2.7b}$$

with the standard normal distribution Φ:

$$\Phi(t) = \frac{1}{\sqrt{2\pi}} \int_{-\infty}^{t} \exp\left(-\frac{x^2}{2}\right) \mathrm{d}x \tag{2.8}$$

With this, the CDF of the lognormal distribution in Eq. (2.7b) can further be expressed as:

$$F(t) = \int_0^t f(x)\,\mathrm{d}x = \frac{1}{2} \cdot \operatorname{erfc}\left(\frac{\mu - \ln(t)}{\sigma\sqrt{2}}\right), \qquad \text{for } t \leq t_{50} \tag{2.9a}$$

$$F(t) = \int_0^t f(x)\,\mathrm{d}x = 1 - \frac{1}{2} \cdot \operatorname{erfc}\left(\frac{\ln(t) - \mu}{\sigma\sqrt{2}}\right), \qquad \text{for } t \geq t_{50} \tag{2.9b}$$

where erfc( ) is the error function complement erfc( ) = 1 − erf( ) and the error function is given as:

$$\operatorname{erf}(t) = \frac{2}{\sqrt{\pi}} \int_0^t \exp\left(-x^2\right) \mathrm{d}x \tag{2.10}$$

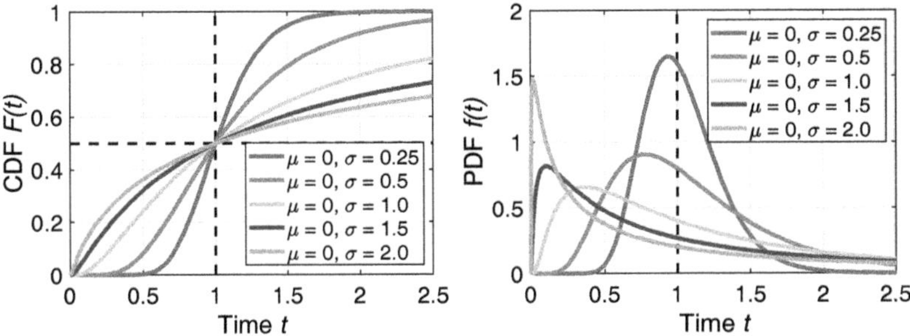

(a) CDF with different shape parameter **(b)** PDF with varying shape parameter

Figure 2.5: Different varying shape parameters of the lognormal distribution. Vertical dashed lines indicate the median, i.e. the t_{50} lifetime, since $t_{50} = \mathrm{e}^{\mu} = \mathrm{e}^{0} = 1$.

The individual formulas of the FR and AFR for the lognormal distribution cannot be simplified as comfortably as for the Weibull distribution and thus are listed in [McP19]. However, the notation in Eq. (2.7b) is especially useful as it relates the lognormal to the standard normal distribution, of which function arguments and values are tabulated as z-values in reference works. Therefore, the values of the lognormal distribution and its inverse function can also be calculated and obtained by hand.

Although the lognormal distribution is always right skewed, it has various shapes that fit many different types of data (see Fig. 2.5). In particular, the multiplicative combination of a large number of independent randomly distributed events is the essential point of the lognormal distribution, as opposed to the normal distribution which considers additive interactions [Bro12, p. 833]. That means, the lognormal distribution models degradation mechanisms which are not localized but act on a macroscopic level, so that all affected regions or components contribute to the degradation or fail of the entire material or system. Typical applications of the lognormal distribution are among others: economic data, response of biological material to stimulus, repair times of equipment, and failure mechanisms like corrosion, wear, fatigue, and creep-induced failure. The lognormal distribution is especially useful when data range over several magnitudes of time. [McP19, Nel82]

2.1.4 Visualization and Analysis

After the mathematical description of failures and their typical statistical distributions have been addressed, one is usually interested in assessing empirical reliability data for appropriate failure distributions and goodness-of-fit (GOF).

On that account, the most common plotting techniques and fit methods for reliability data will be introduced and demonstrated as they are an important foundation in the following chapters.

Probability Plots

For visualization of failure distributions, probability plots are well suited for the visual inspection of goodness-of-fit (GOF) and applicability of the assumed distributions. In order to facilitate visual perception of the matching of empirical data and failure distribution, the already introduced CDF plots (e.g. Fig. 2.2a) are modified in such a way that the distribution function is depicted as a straight line.

This can be achieved by linearizing the respective CDF of the distribution and altering the axis-scaling accordingly. For the Weibull CDF, Eq. (2.6b) transforms into

$$\ln\left\{-\ln\left[1-F(t)\right]\right\} = \beta \cdot \ln\left(\frac{t}{\alpha}\right) \tag{2.11}$$

which can be interpreted as the representation of a linear equation. The quantity Weibit ($\mathrm{Wb} = \ln\{-\ln[1-F(t)]\}$) is artificially introduced for the expression on the left of the equals sign in order to be used as the y-axis of the Weibull probability plot. Its behavior is illustrated in Fig. 2.6a. Of course, to fulfill the general form of the equation of a straight line, the x-axis has to be transformed into a logarithmic time axis. With this representation in Fig. 2.7a, the Weibull distribution parameters α and β can be directly read off the plot. Whereas the scale parameter α or t_{63} is the time value when $\mathrm{Wb} = 0$, the shape parameter β is graphically represented by the slope of the distribution and can be calculated from any two points of the distribution fit line.

The inspection of probability plots helps with the visual perception and comparison of several failure distributions and the applicability of the chosen distribution type to the data. Additionally, it emphasizes the significant tails of the distribution and allowed for graphical fitting in the past when computer-based fitting algorithms and software suites were not available.

For the CDF of the lognormal distribution, Eq. (2.7b) is linearized and gives an expression for the y-axis in an analogue manner, which is called the z-value ($\mathrm{z} = \Phi^{-1}\left[F(t)\right]$):

$$\Phi^{-1}\left[F(t)\right] = \frac{1}{\sigma} \cdot \ln\left(\frac{t}{\exp(\mu)}\right) \tag{2.12}$$

where Φ^{-1} is the inverse standard normal distribution. The x-axis of the lognormal probability plot is a logarithmic time axis and the y-axis is linear in z-values. Unlike for the Weibull plot, the slope of the straight line gives the inverse shape parameter

$1/\sigma$ for the lognormal distribution. Thus historically, σ is usually obtained from the times when the z-values equal 0 and ±1, which approximate to $\sigma \approx \ln(t_{50}) - \ln(t_{16})$ and $\sigma \approx \ln(t_{84}) - \ln(t_{50})$, respectively [McP19]. Additionally, the scale parameter μ is not indicated directly by t_{50}, which is the time when z = 0, but rather by $\mu = \ln(t_{50})$.

As already mentioned in section 2.1.3, Weibull as well as lognormal distributions are flexible and can adapt different shapes. Therefore, it is possible that a set of failure data can be fitted with either distribution. The most pronounced differences are especially noticeable at the tails of the distributions. As shown in Fig. 2.8, when enough data are available, lognormal data will appear concave in a Weibull probability plot and Weibull data will appear convex in a lognormal probability plot. Due to statistical variance as well as other influences like extrinsic failures, the graphical determination of the underlying distribution can be inconclusive. Unfortunately, because of cost restraints like testing time and part quantity it is often not possible to test for these distinguishing distribution characteristics in order to rule out inadequate distributions. As a consequence of these constraints, the choice of a failure distribution should not exclusively be based on empirical data but preferably on theoretical considerations of the failure mode physics. This is especially important, as the fitted distribution is often used to extrapolate the obtained test results by two or three orders of magnitude in time and the failure probability down into ppm (parts per million = 10^{-6}) range. An erroneous assumption and usage of failure distributions will lead to large deviations of lifetime-predictions and possibly also affect reliability-test outcomes. [Str09]

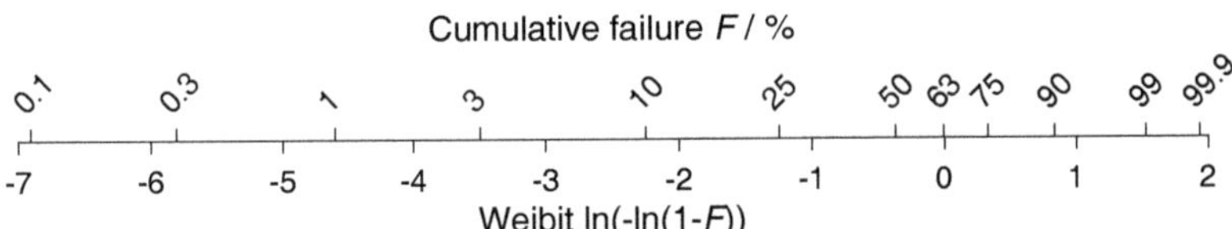

(a) Transformation of the Weibull probability scale into Weibit, see Eq. (2.11)

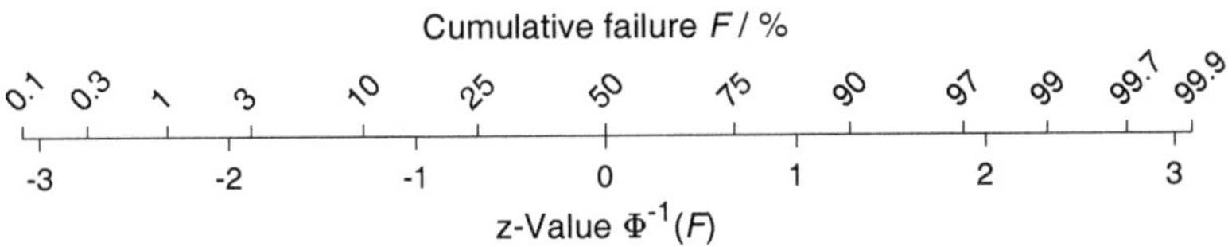

(b) Transformation of the lognormal probability scale into z-values, see Eq. (2.12)

Figure 2.6: Weibit and z-values are the linearized y-axes of the respective probability plots.

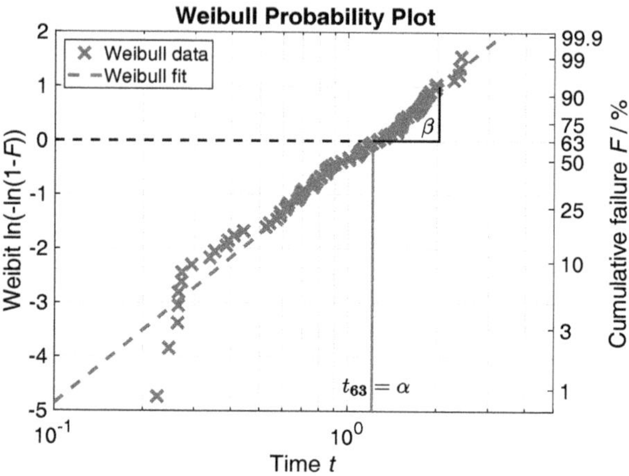

(a) Data from Fig. 2.2a plotted as Weibull probability plot with scaled Weibit axis and logarithmic time axis.

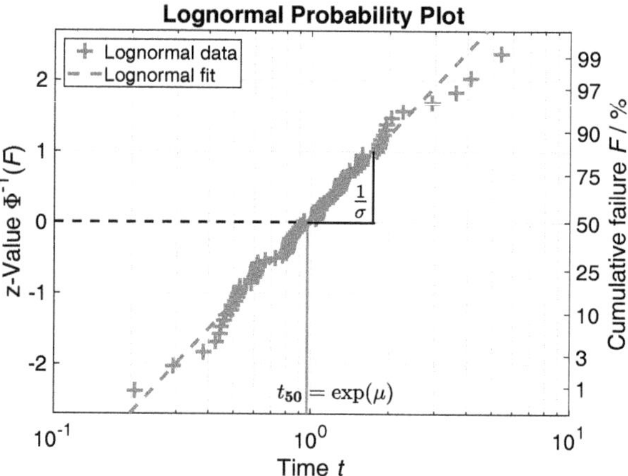

(b) Data plotted as lognormal probability plot with scaled z-value axis and logarithmic time axis.

Figure 2.7: Examples of probability plots with randomly generated failure data of the respective reliability distributions and graphical extraction of failure distribution parameters.

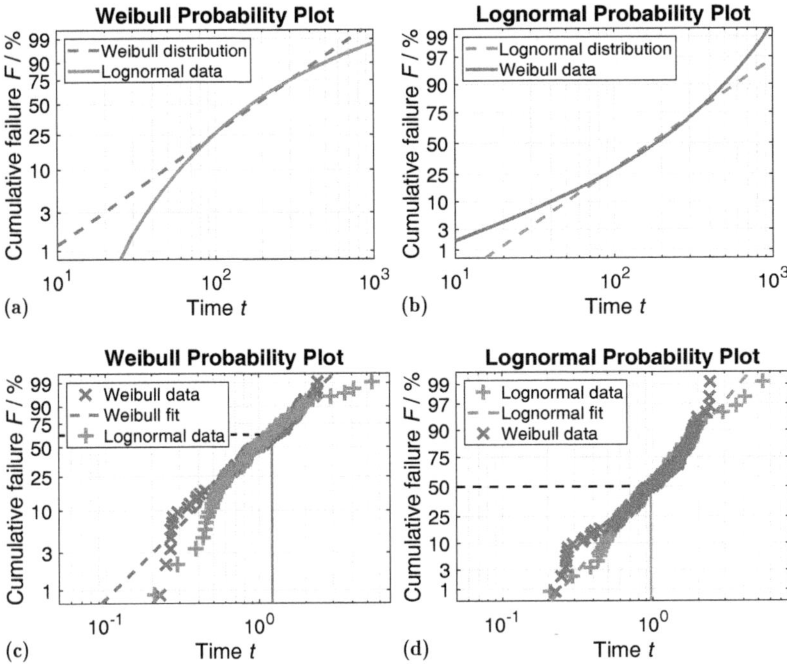

Figure 2.8: Characteristic deviations of failure distributions when plotted in unsuitable probability nets. Subfigures **(a)** and **(c)** show lognormal behavior in a Weibull plot, **(b)** and **(d)** Weibull behavior in a lognormal plot. Whereas **(a)** and **(b)** depict distribution functions, **(c)** and **(d)** show the randomly generated failure data which are also featured in Fig. 2.7.

In times of manual fitting, probability plots were used as the standard evaluation tool for empirical failure data and determination of suitable distributions. Nowadays, fitting is mainly done by algorithms in software suites, but probability plots remain the prime tool for graphical inspection and evaluation of failure data.

Plotting Positions

When it comes to reliability data and plotting them into a probability plot, a choice has to be made where to plot each point on the probability axis, because of the EDF being a step function with a discontinuity at every failure time. The most straightforward approach is to plot the sorted failure times exactly at the read-out times, at which they failed, and at the percentage of cumulated fails:

Each individual i of the total number of specimens n at position i/n. However, by using this plotting position, it is not possible to illustrate the last fail ($i = n$) in the Weibull or lognormal probability plot, because the respective Weibit or z-value would go to infinity. Neither is it possible to plot the first fail, if one decides to use $i-1/n$ as plotting position. The Weibit and z-value would go to negative infinity. This is particularly substantial, if the sample number n is small, as one measurement point cannot be taken into consideration. [Str09, Har84]

For this reason, various new plotting positions have been suggested and advocated in the past that particularly favor certain aspects of distribution fitting or depicting data in favorable manners. Despite this development, the simple plotting position i/n was still used and described by the California Department of Public Works and was therefore called *California method*. Here, a historical overview of different plotting positions is given [Har84]:

$$\text{California method (1923):} \qquad F_i = \frac{i}{n} \tag{2.13a}$$

$$\text{Midpoint position, Hazen (1914):} \qquad F_i = \frac{i - 0.5}{n} \tag{2.13b}$$

$$\text{Mean position, Weibull (1939):} \qquad F_i = \frac{i}{n + 1} \tag{2.13c}$$

$$\text{Modal position, Gumbel (1942):} \qquad F_i = \frac{i - 1}{n - 1} \tag{2.13d}$$

$$\text{Expected value position, Kimball (1946):} \qquad F_i = F\left[\mathrm{E}\left(x_i\right)\right] \tag{2.13e}$$

$$\text{Approx. median position, Lebedev (1952):} \qquad F_i = \frac{i - 0.3}{n + 0.4} \tag{2.13f}$$

$$\text{Generic (symmetric) formula, Blom (1958):} \qquad F_i = \frac{i - \alpha}{n - 2\alpha + 1} \tag{2.13g}$$

The features of these plotting positions are illustrated in Fig. 2.9. Furthermore, there are many arguments for the use of selected plotting positions from Eqq. (2.13) [Har84]:

- Hazen's compromise position Eq. (2.13b) is equivalent to the point midway through the discontinuity from $(i-1)/n$ to i/n and only differs from other plotting positions in the extreme points.
- The mean position Eq. (2.13c) complies with all five of Gumbel's postulates, which plotting positions should satisfy, and gives unbiased estimates.

- Determining the estimated standard deviation is more efficient for Eq. (2.13b) than it is for Eq. (2.13c). Nevertheless, when performing Monte Carlo experiments, the standard deviation is usually underestimated for Eq. (2.13b) and overestimated for Eq. (2.13c).
- The modal position Eq. (2.13d) is the most probable position, therefore called modal. However, it proves less desirable for plotting purposes than Eq. (2.13b) or Eq. (2.13c).
- The expected value position Eq. (2.13e) can give nearly optimal plotting positions. This is especially useful for illustrating the extremes of a failure distribution, due to the fact that it gives unbiased estimates of cumulative probability variables. Having said this, expected values are rather difficult to calculate but can be approximated best with Eq. (2.13g) if no tabulated values are available for certain distributions.
- For normal, lognormal, and in the majority of cases also for Weibull distributions, the approximated median position Eq. (2.13f) is better suited as plotting position than Eq. (2.13b), Eq. (2.13c), or Eq. (2.13e).
- The generic formula Eq. (2.13g) contains all other plotting positions mentioned in Eqq. (2.13) as special cases, but the choice of an optimal value for α is subject to the area of application and in general remains an open debate.

Some influential recommendations for plotting positions in modern references:

- Eq. (2.13b) is used in the software suite MATLAB [Mat18],
- Eq. (2.13g) with $\alpha = 0.375$ is used in the software suite Origin [Ori17],
- Eq. (2.13f) is used in the software suite Minitab [Min19].
- Eq. (2.13f) is suggested by NIST [NIS13] and McPherson [McP19].

With all these various plotting positions available for reliability analysis, the question arises as to whether the differences between individual plotting positions are significant and critical. Figure 2.9 illustrates how different plotting positions have an impact on linearity and overall presentation of failure data in probability plots. As depicted, the actual differences of the illustrated plotting positions are decreasing for an increasing number of sample points independent of the used distribution for the probability plots. Whereas the middle part of the distribution converges, the tails of the distributions exhibit the most noticeable differences. As a rule of thumb, these differences of plotting positions are uncritical beyond the mark of 20 data points [Har84]. It also gets apparent that even these differences are marginal compared to the inherent statistical variance of probability data themselves. [Nel90]

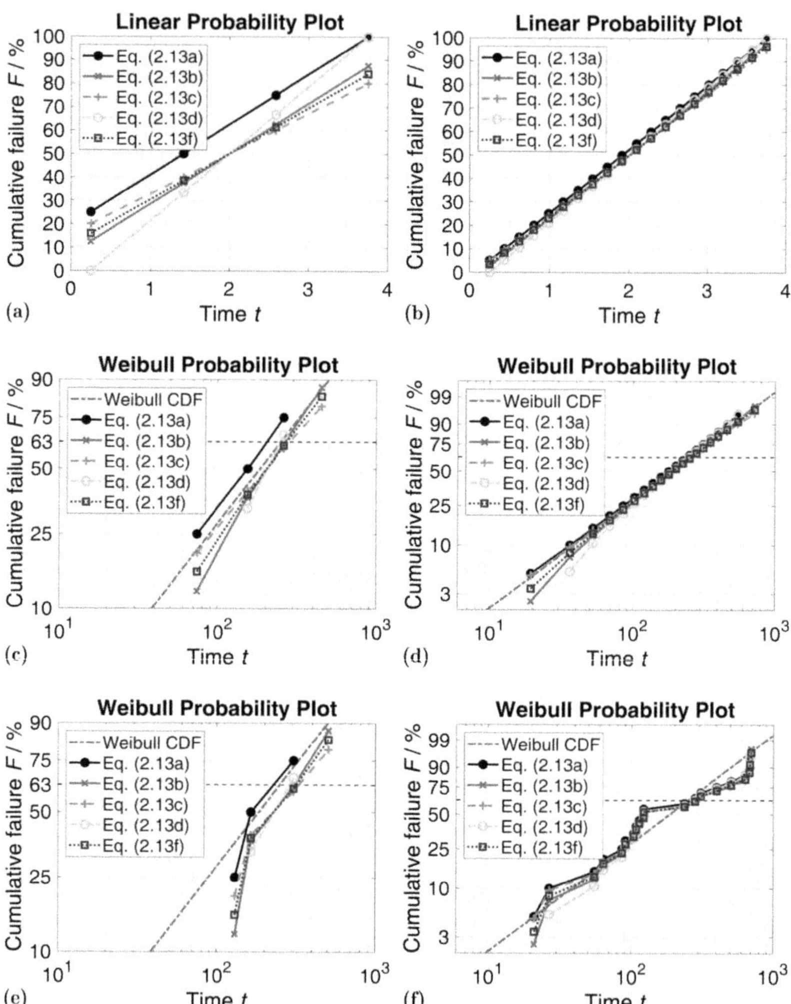

Figure 2.9: Selected plotting positions from Eqq. (2.13) are displayed for purpose of comparability with ideal theoretical failure data in **(a)**, **(b)** linear and **(c)**, **(d)** Weibull probability plots, as well as with **(e)**, **(f)** randomly generated Weibull failure data to illustrate the influence of statistical variance of pure artificial data. Figures on the left are based on 4 data points each, figures on the right on 20 data points each.

Although plotting positions are an important part of reliability analysis and visual assessment of failure data, the choice of the specific plotting position was only critical in times when analysis and distribution fitting was done exclusively by using graphical methods and probability plots. With the assistance of computer programs and advanced fitting algorithms, nowadays the importance of using the best suited plotting positions has receded. [Str09]

Censored Data

Ideally for reliability evaluations, failure or life data of all specimens are available to yield a complete data set. However in practice, reliability data might be incomplete due to either unforeseen events during the test or limitations and restrictions of the test setup, that exclude a number of specimens from the test. These data points are called *censored data* and can not or only partly be considered in the reliability evaluation of the entire population.

This is the case, when specimens fail due to other reasons than the investigated failure mechanism, are removed intermittently or permanently during the test, or the test is still running at the time of a preliminary reliability evaluation. In many cases also premature test ends due to time and cost constraints are reasons for incomplete and therefore censored data [Nel90]. Depending on the type of missing temporal information, censored data are categorized in *left*, *right*, and *interval censored* data. Meaning, that either the part left, right, or a time frame of the time axis, respectively, has not been recorded. The most common types of censored data are shown in Fig. 2.10 and include the following:

Right censored data Type I (time censored) The test ends after a predetermined test period, even if not all specimens have failed by then. These test objects are right censored because the time information about the time of failure on the time axis *right* of the test end is unknown. These data are sometimes called 'truncated' data. The unfailed units are called run-outs, survivors, removals, or suspensions [Nel90].

Typically for stress tests, the test time is fixed, which means that even with careful planning, not all test objects may fail within the scheduled test time. Usually this is the case if there is a defined time budget for testing, if standardized reliability tests are performed, which have fixed time requirements or test times, or due to the measurement equipment's limitation to record only a finite number of data points and therefore having an intrinsic maximum measurement time.

Right censored data Type II (failure censored) The test end is reached when a predetermined number of specimens or proportion of the test population has failed. All remaining test objects are then treated as right censored data.

In practice, this case is not as common as Type I censoring due to the fact that time constraints occur more frequently. However, Type II censoring is more of theoretical interest, because the failure proportion at test end is determined.

Left censored data Data are referred to as left censored, when the failure occurred before (on the *left*) of the first measurement.

This can be the case if no initial integrity test of the test object is performed prior to the stress test and the state of the sample is observed only by the time of the first completed test measurement.

Interval censored data When the specimens cannot be monitored continuously in order to obtain 'exact' failure times but with measurements at discrete points in time, the failure of a specimen is only known to have happened in between two measurements. These data are said to be interval censored.

For certain devices or failure mechanisms, separate equipment has to be used for stressing and testing the devices. These so-called *ex situ* measurements are inherently interval censored, due to read-out times with large intervals. Therefore, *ex situ* data are often referred to as read-out data or grouped data, because interval censored data are sometimes plotted at the time of read-out and, since best practice are log-of-time spread measurement times, multiple fails can accumulate and 'group' at the read-out times. By contrast, *in situ* measurements are considered to give 'exact' failure times, especially if measurement times and read-out intervals are negligibly small compared to the failure times. [Str09]

There are different ways of dealing with censored data and incorporating them into the reliability analysis. Either, censored data are ignored due to the lack of failure information, which however distorts the reliability evaluation, or an attempt can be made to include the time of the last known state of the test object. An especially popular method for considering censored data for reliability analysis and plotting is the Kaplan-Meier method.

Kaplan-Meier Method

The Kaplan-Meier approach can be used for complete or censored data and does not assume any underlying distribution model. Therefore, it is an non-parametric, i.e. empirical, procedure. The estimates obtained by the Kaplan-Meier method are unbiased and also known as product-limit estimates. [Kap58]

Since the plotting problem for the last failure with $F(t) = 1$ in probability plots still exists, the *modified* Kaplan-Meier estimates are often used for probability plotting. For a total number of n samples on the test, the running times are arranged from t_1 to t_n. These times include actual failure times as well as last observation times

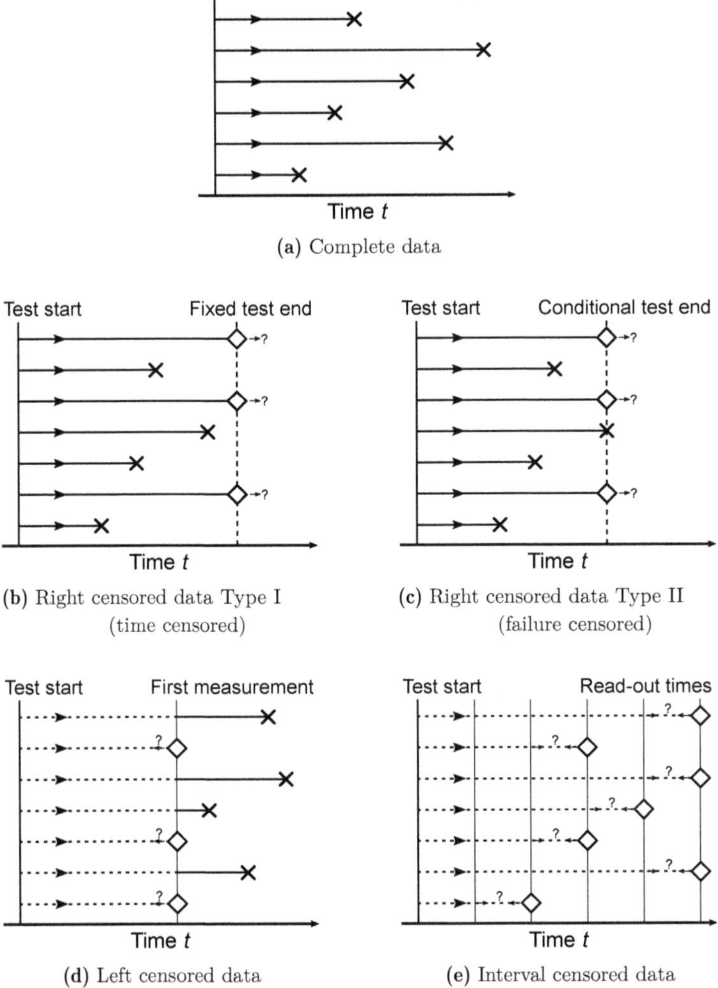

(a) Complete data

(b) Right censored data Type I (time censored)

(c) Right censored data Type II (failure censored)

(d) Left censored data

(e) Interval censored data

Figure 2.10: Different types of censored data. Exact failure times are marked with crosses $\times$, censored data with diamonds $\diamond$ and time spans of unknown state with question marks ?. Solid horizontal lines denote time spans of continuous monitoring of the specimens, dashed horizontal lines represent intervals of unknown failure state of the test objects.

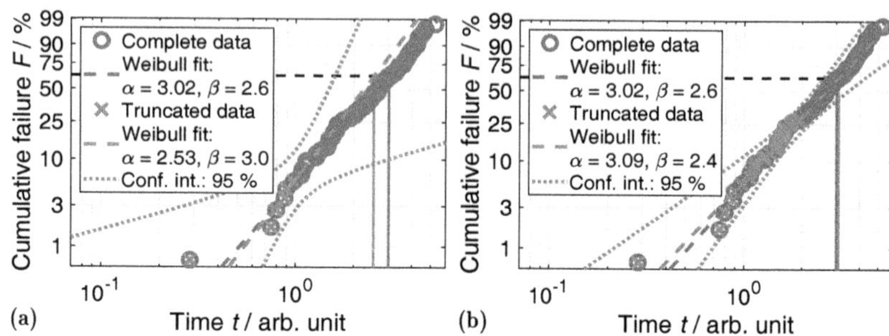

Figure 2.11: Complete randomly generated reliability data displayed with blue circles ○ and derived truncated data, i.e. right censored data Type I, plotted with red crosses × using modified Kaplan-Meier estimates. With Weibull fits and the 95 % confidence intervals, reliability predictions of the sample population can be made based on the censored data, despite of the failure data being truncated to less than **(a)** 20 % and **(b)** 35 % of the last failure time of the complete data.

of censored specimens. From the total set of indices from 1 to n of all samples, only those that correspond to actual failure times are used for calculating the CDF estimates. The used indices might therefore be a discontinuous series of increasing numbers:

$$\hat{F}(t_i) = 1 - \frac{n+0.7}{n+0.4} \prod_{t_j \leq t_i} \frac{n-j+0.7}{n-j+1.7} \tag{2.14}$$

These CDF estimates, indicated by the ^ symbol, equate to the approximated median ranks Eq. (2.13f) for plotting Type I censored data. [NIS13]

Due to its universality to deal with censored and complete reliability data, the modified Kaplan-Meier method is implemented throughout this work for all probability plots. This also applies to the example in Fig. 2.11, in which a reliability analysis called *sudden-death*-method is demonstrated. Suppose, a company launches a new product and after several months is faced with an unusually high number of returns due to product failure. Thus, the company wants to evaluate the reliability properties of the remaining product population that is still in the hands of the customers in order to plan its future course of action. In such a case where only incomplete and truncated data are available, the existing failure times can be plotted using modified Kaplan-Meier estimates and a preliminary reliability assessment of the sample population can be made with some degree of uncertainty, which can be expressed with confidence intervals.

After this outline of fundamental plotting techniques necessary for reliability analysis, the following is a brief orientation of the most important analysis methodologies on fitting distributions and statistical comparative tests, which are also used throughout this work.

Maximum Likelihood Estimation

The maximum likelihood estimation (MLE) is a method for estimating the most likely parameters of a distribution for given data points and thus fitting them. It is a totally analytical maximization procedure, and as it can be used for any form of censored data, it is especially suited for distribution fitting of reliability data. [NIS13]

Generally, for a random variable X which is distributed according to a PDF $f(\,)$ with parameter vector $\boldsymbol{\theta} = (\theta_1, \theta_2, ..., \theta_k)$ — for Weibull distributions these correspond to the parameters (α, β), for lognormal distributions to (μ, σ), respectively — the total probability for n samples with the results $x_1, x_2, ..., x_n$ is obtained by the product of the individual probabilities [Rel15]:

$$f(x_1, x_2, ..., x_n; \theta_1, \theta_2, ..., \theta_k) = \prod_{i=1}^{n} f(x_i; \theta_1, \theta_2, ..., \theta_k) \tag{2.15}$$

Focusing on the set of unknown parameter θ_j for the given results x_i, which are to be fitted, one defines the likelihood function L:

$$L(\theta_1, \theta_2, ..., \theta_k \mid x_1, x_2, ..., x_n) = \prod_{i=1}^{n} f(x_i; \theta_1, \theta_2, ..., \theta_k) \tag{2.16}$$

The objective is to maximize the likelihood L fitting parameters θ_j for the given results x_i. Due to the difficulties of differentiating the product of distribution functions, one can take advantage of the logarithmic calculation rules by first logarithmizing the likelihood function L:

$$\mathcal{L}(\boldsymbol{\theta}) = \ln\left(L(\boldsymbol{\theta})\right) = \ln\left(\prod_{i=1}^{n} f(x_i; \theta_1, \theta_2, ..., \theta_k)\right) = \sum_{i=1}^{n} \ln f(x_i; \theta_1, \theta_2, ..., \theta_k) \tag{2.17}$$

The newly obtained log-likelihood function $\mathcal{L}$ is therefore better suited for optimization towards the maximum regarding $\theta_1, \theta_2, ..., \theta_k$:

$$\frac{\partial \mathcal{L}(\boldsymbol{\theta})}{\partial \theta_j} \stackrel{!}{=} 0\,, \text{ for } j = 1, 2, ..., k \tag{2.18}$$

The subsequent question is: What advantages and deficiencies does the MLE have for the purpose of distribution fitting compared to a linear least squares fit on the already discussed probability plots?

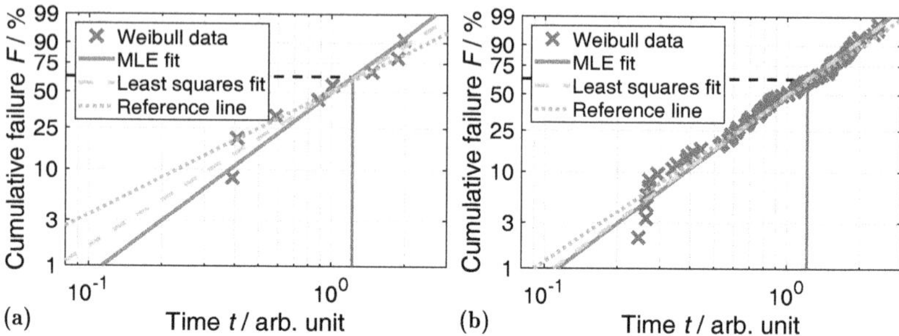

Figure 2.12: Randomly generated data with sample size **(a)** $n = 8$ and **(b)** $n = 80$ are fitted likewise with MLE and least squares method. Also a reference line is shown, which intersects the EDF at the 25th and 75th percentile.

To begin with, the maximum likelihood estimation fits all common distributions and is usually the statistically most efficient method for estimating certain parameters. Furthermore, the resulting maximum likelihood estimates $\hat{\boldsymbol{\theta}}$ become unbiased for increasing number of samples, with the smallest variance of all estimators of this kind. MLE is versatile and can be used for the determination of confidence intervals, t-tests, and hypothesis testing in general. Additionally, it is independent of the formerly discussed plotting positions on the y-axis. While this method is very precise for large samples, the results can be strongly biased for small samples ($n \lesssim 5$). Also note that MLE often requires special software to solve nonlinear equations systems. [NIS13]

It should be noted that the fitted distributions by means of MLE are independent of any ranks or plotting positions chosen for probability plots. Thus, the plotted failure points and the plotted distribution line can seem to not always match together perfectly, which can also be seen in Fig. 2.9 and should not be reason for confusion or doubt in any of these techniques. [Rel15]

On the contrary, linear *least squares* fits are easy to carry out, also possibly by hand. Nevertheless, in cases of a large variance of the error, least squares fits can give high correlation coefficients even if there is no existing correlation of the data. Also, the susceptibility to outliers is very high, especially if they are at the beginning or end of the distribution. This is particularly unfavorable, because these points are by nature less constrained in their plotting positions as the central points in probability diagrams are. As a last point, with linear least squares fits of distributions, confidence intervals can only be calculated to a limited extent, since after linearization of the distribution equations, the used regression models are no longer valid. [Str09]

In Fig. 2.12, two sets of randomly generated data are fitted with MLE and least squares fit for comparison. Additionally, a so-called *reference line* is plotted, which is supposed to be a guide to the eye. They are found in various software suites but are not related to the presented fit methods in any way.

For the reasons stated above, all major software suites for reliability analysis use the maximum likelihood estimation as default distribution fitting method. However, for the specific use of comparing several probability distributions with a goodness-of-fit measure for their applicability to measured reliability data, the MLE's figures of merit are less suitable. This is the domain of other statistical tests, like the following one.

Anderson-Darling Test

The Anderson-Darling (AD) test, proposed in [And52, And54], is a statistical test to determine how far a given frequency distribution deviates from an assumed probability distribution. It can not only be used as a goodness-of-fit (GOF) indicator for the fit of a single distribution but also for comparing different distributions of the same fit data. The Anderson-Darling test is not the only test that can achieve this but is with others, like chi-square, Kolmogorov-Smirnov, or Cramér-von Mises, among of the most powerful statistical tests for this application [NIS13, Ste74].

In principle, the test calculates the weighted distance A of the empirical CDF (i.e. EDF) F_n from the hypothetical CDF F. With $w = [F\,(1-F)]^{-1}$, the margins of the distribution are particularly strongly weighted [And52, And54]:

$$A = n \int_{-\infty}^{\infty} \frac{[F_n(x) - F(x)]^2}{F(x)\ [1 - F(x)]} \, \mathrm{d}F(x) \tag{2.19}$$

More precisely, the AD test generates a figure of merit A^2, which is compared with a critical value k, which depends on type of distribution F, number of samples n, and significance level p. These critical values have been tabulated in literature for various cases and distributions and are usually integrated in standard software suites:

$$A^2 = -n - \sum_{i=1}^{n} \frac{(2i-1)}{n} \left[\ln F(x_i) + \ln\left(1 - F(x_{n+1-i})\right)\right] \tag{2.20}$$

If $A^2 > k$, the null hypothesis H_0 ("The frequency distribution of the data in the sample corresponds to the given hypothetical probability distribution.") is rejected. The value of p at which $k(F, p, n) = A^2$, which is the upper limit where the null hypothesis would not be discarded, can be used as a GOF output for comparison purposes between different probability distributions fitted to the same reliability data. [And54, DAg86, NIS13]

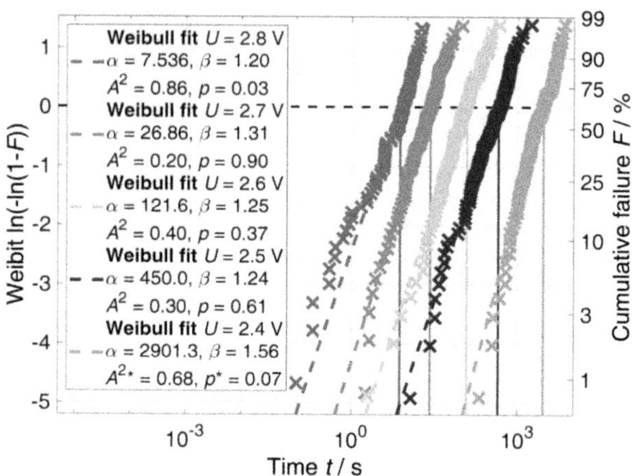

(a) Five time-dependent dielectric breakdown (TDDB) Weibull distributions are measured with different stress voltages. The critical value k of the AD test for the given sample number of these distributions and a standard 5 % significance level is $k = 0.75$, at which the null hypothesis H_0 is rejected. An asterisk * marks data sets containing censored data, whose AD statistics are inaccurate and might not be conclusive.

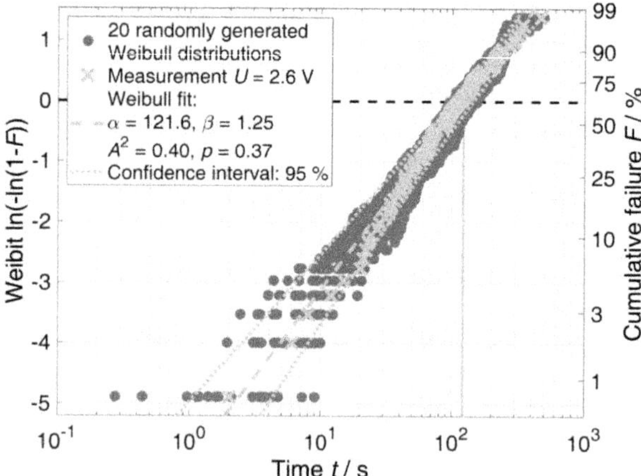

(b) A single experimental failure distribution from Fig. 2.13a with a 95 % confidence interval is underlayed with 20 randomly generated Weibull distributions, generated with the distribution parameters given from the MLE fit of the experimental data.

Figure 2.13: An exemplary voltage stress series of Weibull distributed TDDB failure data to convey some intuition for statistical variance and AD test results.

To facilitate a perception of GOF of genuine experimental TDDB failure data, Fig. 2.13 illustrates different Weibull failure distributions and their AD statistics. In particular, Fig. 2.13b depicts an experimentally measured failure distribution with the 95 % confidence interval, which is compared with 20 randomly generated Weibull distributions stemming from the same distribution parameters to show the intrinsic statistical variance of failure data. That is because statistically, the confidence interval at a level of 95 % of only 1 in 20 EDFs is likely to not cover the 'true' underlying failure distribution.

As further extensive literature on goodness-of-fit techniques, [DAg86] is recommended as reading material. This concludes the introductory chapter on fundamental failure distribution mathematics, plotting, and statistics for the purpose of this work. With this knowledge, the conducted reliability analysis and predictions in the following chapters are sufficiently comprehensible.

2.2 Reliability Methodology

A broader overview of this subject apart from the already recommended literature is given in the "Handbook for Robustness Validation of Semiconductor Devices in Automotive Applications" of the German Electrical and Electronic Manufacturers' Association (ZVEI) [ZVE15]. The statements and conclusions presented here are assembled from individual discussions with reliability experts and industry knowledge. They can also be found to some extent in the standard references already mentioned.

Starting with the general conception of lifetime extrapolation and the required steps for deriving a lifetime prediction under use conditions, potential pitfalls and challenges during implementation are also addressed and discussed. Subsequently, a general guidance is given on special considerations for data analysis and evaluation.

2.2.1 Reliability Testing and Lifetime Predictions

Although all three regions of the bathtub curve in section 2.1.2 are important for the dimensioning of reliability and robustness margins, lifetime extrapolation is primarily conducted with the intrinsic wear-out mechanisms and failure rate in mind. EFR and IFR are usually considered separately in reliability and robustness validation. Hence, this work focuses explicitly on the intrinsic wear-out mechanisms for reliability qualification.

The common process of deriving a lifetime prediction for a product under given requirements and use conditions is depicted in Fig. 2.14:

1. Empirical failure data have to be gathered individually on any relevant failure mechanism. Advantageously, this is done at different stress levels of the known and relevant stressors of each specific failure mechanism, like temperature, voltage, relative humidity, etc. The acquired CDF fit data also contain GOF information that can be used to calculate confidence intervals for the later needed scale and shape parameter extrapolations.

2. The obtained failure distributions under stress conditions are then extrapolated with an applicable acceleration model to normal operating or use conditions, usually stated by the customer. This extrapolation to use conditions is performed using the respective CDFs' characteristic scale parameters $\alpha = t_{63}$ or $\mu = \ln(t_{50})$, or any other $t_{F/\%}$ failure quantile. If several stressors have been used to accelerate the failure distribution, the respective acceleration factors have to be applied simultaneously. Of course, this AF extrapolation is afflicted with uncertainties and errors, therefore prediction intervals should be included in the subsequent calculations.

3. Since test structures usually differ from the final products, the failure behavior has to be scaled to the respective dimensions, e.g. gate area, line cross section, or line length of the actual semiconductor chip design. As example, Poisson area scaling is depicted in Fig. 2.14 as vertical shift of the distribution on an area-normalized Weibit axis, which will be explained in more detail in section 3.3.

4. After deriving the CDF of the considered failure mechanism for the product under use conditions, this distribution is extrapolated statistically to lower failure quantiles with respect to the determined confidence intervals of the distribution fitting and the stress extrapolation to use conditions.

5. & 6. To the required target failure quantile an expected lifetime with confidence intervals can be obtained for the product under use conditions.

It is evident from Fig. 2.14 that a robust lifetime extrapolation according to this procedure is only valid if each acceleration and extrapolation step is a uniform transformation that performs parallel shifts of the failure distribution in the probability plot. A change in the Weibull slope β over any stress range that is regarded for the extrapolation would significantly alter the value of the expected final lifetime. In addition, sample sizes and stress levels for the reliability measurements should be chosen wisely to obtain representative and robust data to work with.

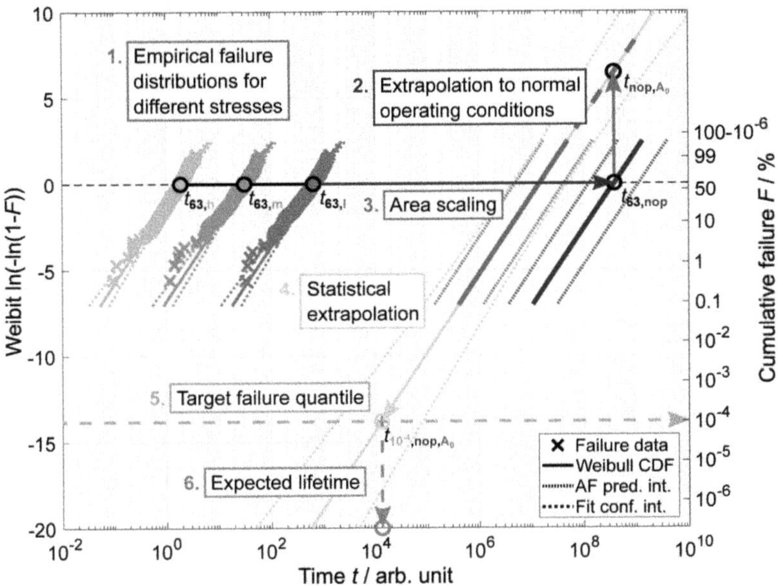

Figure 2.14: Method of deriving a lifetime prediction for a product under use conditions from initially conducted stress test measurements. First, the fitted failure distributions are extrapolated to normal operating stress conditions via a stress acceleration model and scaled to the required area or cross section. With subsequent statistical extrapolation to the target failure quantile, the expected lifetime can be read off the time-axis.

2.2.2 Accelerated Lifetime Testing

While the previously described statistical extrapolation in steps 4 to 6 of the lifetime prediction procedure in Fig. 2.14 are merely of statistical nature, the extrapolation to use conditions in step 2 requires profound physical considerations. Beyond the justified choice of an appropriate and verified acceleration model for the calculation of the acceleration factor, the choice of testing method and the extend of stress acceleration is also a vital one.

As depicted in Fig. 2.15, the example of an automotive application requires a expected product lifetime at specified use conditions, which could be given as an mission profile, is set to be 10 years. Standard matured accelerated lifetime testing (ALT) methods are usually accelerated by a factor of about 10 to 100 (see Eq. (2.5)).

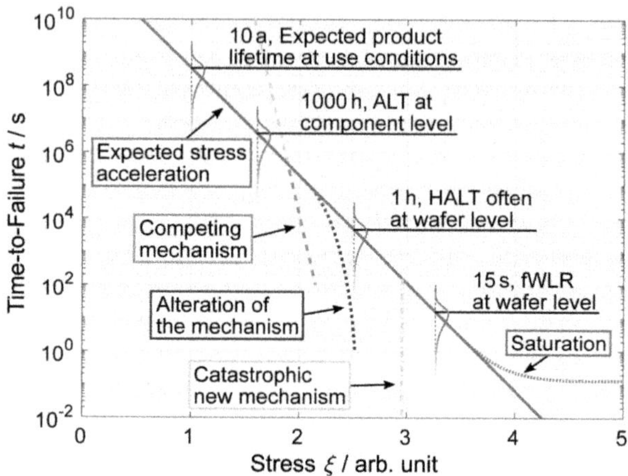

Figure 2.15: Stress acceleration behavior for a given failure mechanism that allows to extrapolate failure distributions under stress conditions to use conditions. However, the extrapolation of the depicted t_{63} test times of accelerated lifetime testing (ALT), highly accelerated lifetime testing (HALT), and fast wafer level reliability (fWLR) measurements can be affected by various emerging mechanisms.

An example for the stated test time of 1000 h is the high temperature operating life (HTOL) test in the requirements of the Automotive Electronics Council (AEC) AEC-Q100 [Aut14], which is usually performed on the finished product or component, meaning a fully designed, fabricated, and packaged semiconductor chip on a test circuit board.

Due to the long test times in the order of months necessary for the final reliability qualification, reliability assessment during product and technology development has the need for faster and therefore highly accelerated lifetime testing (HALT). Due to the lack of spare time and a final product, HALT is usually conducted on diced chips, so-called *dies*, or directly on wafer level. This includes testing on final, designed chips, specialized test structures and chips or individual devices during technology development. Stress and failure times are often in the duration of hours for each device under test (DUT), that can be tested in serial or parallel. This results in acceleration factors in the order of 10^5 in respect to the use conditions which is often achieved by the acceleration of the failure mechanism by multiple stressors.

Special cases in which short testing times are imperative, such as production in-line testing for reliability surveillance of each and every wafer or die, are the domain of testing methods described as fast wafer level reliability (fWLR). This is accomplished by extremely accelerated or progressive stress conditions with an acceleration factor of up to 10^7 and more. Another example is the application of fWLR as pre-characterization tests for other HALT procedures and can be found in the Joint Electron Device Engineering Council (JEDEC) standard JESD92 [JED03] as ramp-stress testing, lasting only seconds.

This extremely high error acceleration also poses dangers for the integrity and applicability of the final lifetime extrapolation to the designated use conditions. Figure 2.15 illustrates how other mechanisms and alteration of the same failure mechanism can interfere with the assumed acceleration model, which then results in a strongly deviating lifetime prediction.

Although one aims to look at individual failure mechanisms, it is not always possible to isolate and expose them properly. A competing mechanism that is marginal and negligible at use conditions could be accelerated faster by the same stressor as the failure mechanism under investigation and at some stress level emerge and predominate the observed failure behavior.

Another possible interference of the lifetime extrapolation is an alteration of the examined failure mechanism. As the mechanism blends over into another type of the same mechanism with altered behavior, detecting this transition is usually more difficult. Also finding the transition stress level and identifying both stress acceleration model components is a task considered not possible in a straightforward way [Str09] but is being attempted to be implemented as unifying or combined model for certain failure mechanisms, like TDDB [Hu99, McP00].

Therefore, it does not seem possible or reasonable to increase the stress beyond the occurrence of these mechanisms. Either exclusively competitive mechanisms would be tested, which are regarded as marginal under use conditions, or in case of unknown processes the extrapolation based on the known acceleration model equations is no longer guaranteed.

The occurrence of a catastrophic mechanism at some highly accelerated stress level is a certain terminal event as any further stress acceleration is not possible beyond this point. Weak points in design or material choice are often revealed which are only relevant at these high stress levels but not at use conditions, yet they impede further stress time reduction. Examples for these catastrophic failures are the melting temperatures of packaging materials or solder connections, also disruptive discharge voltages over air or between test leads.

A different, but not as disruptive interference as a new catastrophic failure behavior is a possible lifetime saturation on the lower limit of the time-axis, where a higher stress level does not reduce the life and test time anymore. This could be due to stress time domains being too short for the failure mechanism dynamic to unfold, like molecule diffusion and drift processes.

All these effects are particularly severe for the stress acceleration calculations if they haven't been identified for the failure mechanism or the device yet, or if they are not

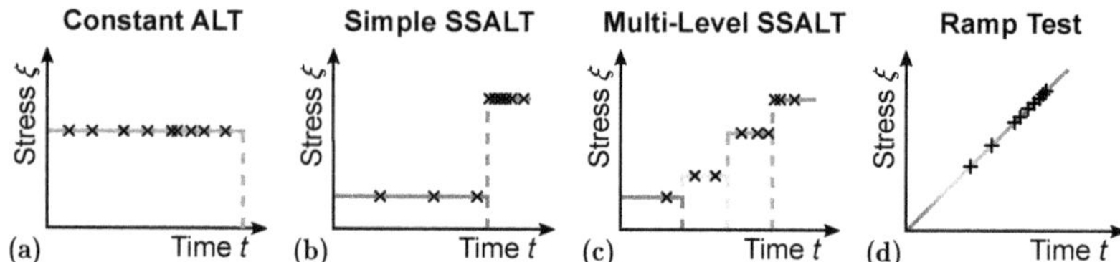

Figure 2.16: Variety of stress methods with different properties: **(a)** Constant accelerated lifetime testing (ALT) features a single constant stress level, whereas **(b)** simple and **(c)** multi-level step-stress accelerated life testing (SSALT) are composed of a piecewise succession of constant stress steps. An even more elaborate method of stress testing is **(d)** a continuously increasing ramp-stress test.

noticed by the reliability engineer. The difficulty in detecting diverging behavior of the chosen acceleration model is that the mechanisms are often only noticeable if stress tests have been conducted over several orders of magnitude of TTF, reaching as close as possible to the stress levels and lifetimes of the use condition. Also note that the TTF regions where deviating behaviors occur in Fig. 2.15 are chosen arbitrarily and could also be present in any other stress regimes.

The most commonly applied test types and methods to achieve stress acceleration are presented in Fig. 2.16, each having different advantages and characteristics that make them particularly useful for specific test applications. There is the previously mentioned ALT method (see Fig. 2.16a), commonly referring to a constant stress level, that is higher than the specified use conditions. ALT the most basic and established stress method and widely used in reliability qualification processes and required in qualification standards, because it primarily produces the same error mechanism. Multiple constant ALT measurements provide the possibility of determining the acceleration factor and the physical failure model in a simple and accessible way.

A disadvantage of measurements with constant stress is the long measurement time if the stress level is set too low. This often leads to the situation that not all components fail in the intended stress time or the measurement has to be aborted for a lack of available time. This circumstance can be avoided by a SSALT measurement, in which the DUTs are stressed by starting with a low stress level, but increasing it either once (see Fig. 2.16b) or consecutively during the measurement (see Fig. 2.16c) in order to cause even the strongest components to fail. Intensive test planning is required to optimally distribute the stress steps and their levels within the targeted test time. This requires knowledge about the acceleration model and physical understanding of the failure mechanism. A distinction is often made in the terminology as to whether the first stress level is as low as the operational stress and is thus only *partially accelerated*, or higher and is therefore called *fully accelerated*.

A further development of SSALT is the ramp-stress as depicted in Fig. 2.16d. Here, the stress level is increased continuously and progressively to cause all specimens to fail in a shorter and predetermined time. For this, in-depth knowledge of the acceleration models for the failure mechanism is necessary in order to integrate the individual stress elements into a total stress and to be able to determine the effective failure times accordingly.

This list is supplemented by stress methods that define more particular stress sequences. One of these is cyclical stress testing (CST), which is used mainly for testing mechanical failure mechanisms under temperature cycling. CST will also be used in chapter 4, and will be discussed in more detail later on. In addition, there are other stress methods which are not part of the standard repertoire for reliability qualification. These include randomly varying stress levels, so-called *stochastic loads*, for which a specific distribution or autocorrelation may be used. On the other hand, stress tests can also be performed with non-repeating stress patterns. [Nel90]

As has been outlined, the use of accelerated testing must be conducted with caution and the advantages and disadvantages of certain stress test techniques must be weighed before use. Once data have been collected, they must also be processed and evaluated in a thorough and structured manner, for which a few important guidelines and suggestions will be given in the following.

2.2.3 Data Analysis and Evaluation

For a robust processing and evaluation of the recorded raw data, the following considerations may prove beneficial for data analysis. As a first step, it is important to define adequate failure criteria and to check how these thresholds reflect the actual failure behavior of individual measurements and DUTs. While some failure mechanisms cause gradual degradation that affects performance, others tend to cause sudden and catastrophic failures. Therefore, exceeding the specification is already considered a failure in semiconductor devices, as this may lead to serious problems in digital circuits. Subsequently, devices that have not failed or were already defective before the stress measurement must either be classified into the various censoring types of section 2.1.4 or sorted out as production yield loss. For this purpose, it is helpful to check the integrity and function of the DUTs before and after stress measurements.

The failure events of the individual measurements are subsequently combined and result in accumulated lifetime data of the stress test. To further process these data, it is necessary to sort them into a suitable order. Depending on the analysis objective, it might be required to sort failure times in ascending order to plot them as EDF in a probability diagram or, for example, by location of failure occurrence in order to facilitate the identification of design or production related weak points.

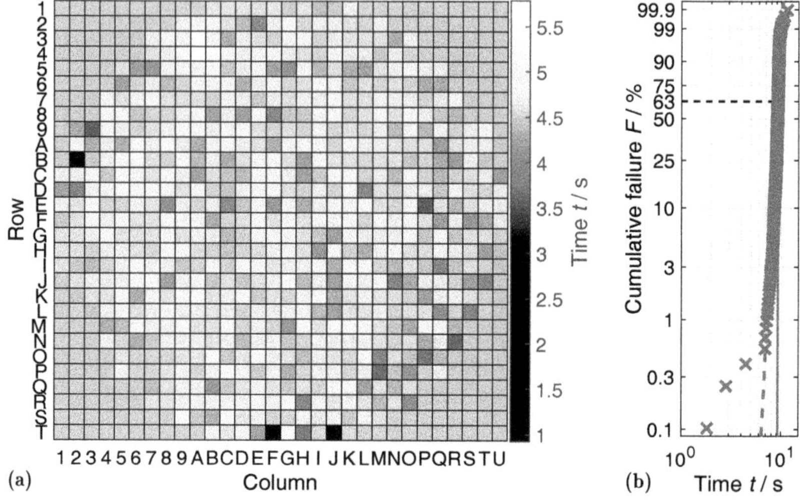

Figure 2.17: **(a)** Heat maps visualize the measured TDDB failure times in relation to their location on a wafer map to enable the detection of location-dependent effects such as failure hotspots, also called 'bad neighborhoods'. **(b)** Weibull probability plot of the same failure data. The first 3 failures are of extrinsic nature and correspond to the dark colored edge dies in the heat map. The gradual flattening of the distribution towards higher failure fractions with a slight bend at the top is an indication of the extent of oxide-thickness nonuniformity across the wafer.

Once the failure data are accumulated and sorted, they can be analyzed and checked for sanity using a wide variety of methods. This is particularly important to determine whether there are external influences on the assumed uniform distribution of the failure data. One example is the heat map in Fig. 2.17 which can be used to detect deviating areas, radial distributions, or fracture edges. In addition, it is possible to test for various correlations. Often, the lifetime is plotted against the sequence of measurements, as shown in Fig. 2.18, in order to identify any temporal characteristics or trends during a serial measurement of DUTs. Other popular analysis methods include histograms, probability plots, statistical tests, and others, like discussed in section 2.1.4.

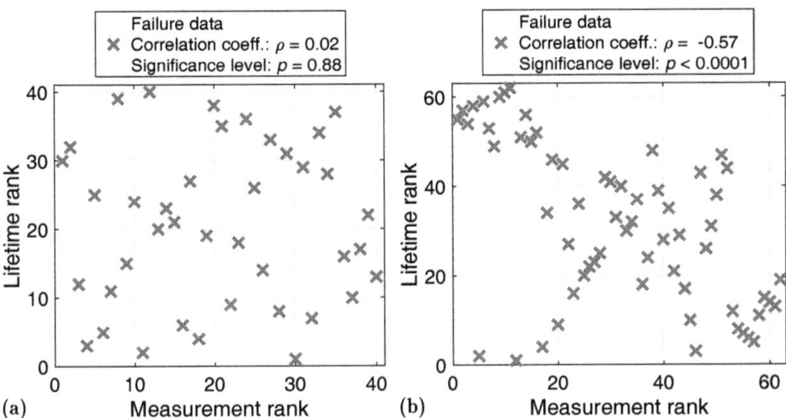

Figure 2.18: Correlation plots expose relationships of failure times to the sequence in which DUTs are measured. **(a)** Uncorrelated failure data. **(b)** Data of a measurement that was impaired from measurement rank 21 on due to increased mechanical pressure on the contact pads (see Fig. 3.4b).

Finally, there is the decision on which failure distribution to use and the choice of an applicable stress acceleration model, which will be discussed in section 3.3. This concludes the general overview of evaluation and analysis of reliability data and is continued by the introduction of a concrete exemplary failure mechanism, which will be utilized for the rest of this work. During this course, the methods of lifetime prediction and reliability assessment presented here will be elaborated and demonstrated with specific data and models.

3 The Failure Mechanism of Time-Dependent Dielectric Breakdown

Subsequent to the presentation of reliability methodology and lifetime predictions, these methods will be illustrated by using the failure mechanism time-dependent dielectric breakdown (TDDB) as a practical example. After the fundamental failure physics of TDDB has been considered, an overview is given of the test structures, devices and methods used throughout this work. The chapter is concluded by the introduction and assessment of different voltage acceleration models, temperature acceleration, and area scaling and insight is provided into industry reliability qualification methods using this mechanism.

3.1 Physics of Dielectric Breakdown

Dielectric materials are subject to degradation under electrical and thermal stress, which affects their electrically insulating properties. When degradation of the material is advanced and the damage is too severe, the dielectric layer will break down.

Dielectric breakdown can occur in different locations inside a semiconductor chip and is therefore divided into groups according to their appearances. As illustrated in Fig. 3.1, TDDB is usually distinguished between front end, back end [JED16] and sometimes also middle of line [Wu20] (FEOL, BEOL, MOL) TDDB. This classification depends on the occurrence and the material properties of the dielectrics that break down as well as the materials of the conducting electrodes. This work focuses on the FEOL TDDB, as this is often the most critical type of dielectric breakdown concerning the expected lifetime, mainly due to the importance of dielectric integrity and thickness of this particular insulating layer.

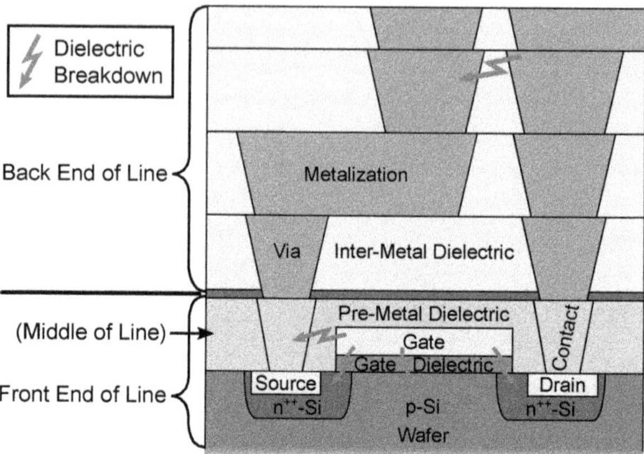

Figure 3.1: Schematic of a highly integrated metal–oxide–semiconductor field-effect transistor (MOSFET) with highlighted locations of possible time-dependent dielectric breakdown (TDDB). They are differentiated into front end of line (FEOL) TDDB in the gate dielectric, middle of line (MOL) TDDB in the pre-metal dielectric, and back end of line (BEOL) TDDB in the inter-metal dielectric.

Yet for this gate dielectric, different materials are used for advancing technology nodes. The originally used silicon dioxide (SiO_2) has been replaced by high-k dielectrics for leading edge technologies.

Electrical stress causes degradation of the dielectric between the gate electrode of the transistor and the bulk silicon of the wafer. While voltage is applied to the contacts of the dielectric, causing an electric field to build up, bond breaking among the insulator atoms appears continuously throughout the dielectric film and generates defect trap states. Above a critical defect density, a conductive path is stochastically formed through the insulator layer, initiating the breakdown event. This allows a current to flow through the dielectric and causes Joule heating. With enough current, the temperature increases rapidly, destroys the insulator layer, and melts adjacent silicon or metal contacts. The molten conducting material can form a permanent short circuit path between the bulk and gate electrode. [McP19, Str09, Ala02a, Ala02b]

The generation of defect traps and the formation of a conducting path is described by the percolation theory. Figure 3.2 illustrates that a conducting path of trap states is formed stochastically and can be composed of a varying number of overlapping defects' spheres of influence [Som02, Ala02b, McP12].

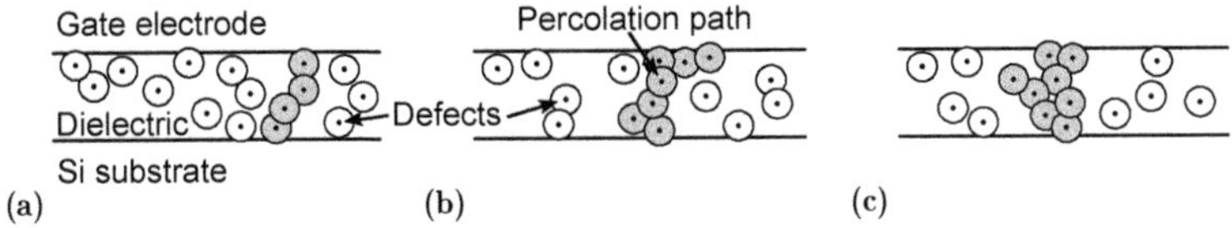

Figure 3.2: Percolation theory describes TDDB behavior microscopically as random generation of defects in the dielectric layer, forming a percolation path at a device specific critical defect density, that can consist of different quantities of defect traps states. This example of a dielectric layer has a thickness of about four times the defect diameter. These random breakdown paths consist of **(a)** 4, **(b)** 7, and **(c)** 8 defects, which result in a different conductance for each path.

As topographical and electrical measurements with conductive atomic force microscopy show, TDDB events are locally confined to areas of approximately 10–100 nm^2, but the morphological and electrical effects can propagate to neighboring areas and influence larger regions of the dielectric [Por03, Eft07].

These extremely localized conducting breakdown paths are the origin of TDDB and cause the entire capacitor or transistor to fail. For that reason, TDDB is a showcase example for weakest-link type failure mechanisms that are well suited to be fitted by Weibull distributions [McP19]. Experimental investigations and the results confirm this theoretical consideration as well [Deg00]. It is also concluded from percolation theory that the lifetime as well as the Weibull slope is decreased for thinner dielectrics [Som02, Wu02]. This can be explained by the statistical probability that the formation of a breakdown path in thinner layers is achieved more likely in shorter time frames and with less temporal variations than in thicker dielectrics. Whether defect generation occurs uniformly or with partial clustering is subject of current research [Agu19].

Furthermore, breakdown events can unfold in very short time frames, which are in the order of nanoseconds [Str09], thus conventional measurement equipment for reliability testing cannot resolve the sudden current increase and only display adjacent measurement points with a current jump. This results in a distinction of different types of breakdowns. Due to their characteristics, TDDB is distinguished between a typical *hard breakdown* with a current jump of at least a factor 10 and a *soft breakdown* that is often observed in extremely thin oxides, exhibits random telegraph noise or higher leakage current before breakdown, and is more difficult to detect [JED03]. The difference in behavior can be attributed to a required critical electric power P_{crit} to reach the necessary temperature for a catastrophic breakdown event, which is described above and can be discriminated from the soft breakdown by their post-breakdown I–U characteristics [Ala02a].

Whether P_{crit} is achieved in a breakdown event or not also depends on the impedance and capacitance of the measurement setup [Jac97, Lin01]. Without this critical power so-called *progressive breakdowns* can accumulate, which keep increasing the leakage current but do not necessarily result in a failure of the device, but interestingly, the failure times from the onset of these first breakdowns as well as the final failures are both Weibull distributed [Wu09c, Deg00].

3.2 Test Structures, Devices and Methodology

In order to utilize the failure behavior of FEOL TDDB as a practical example for this work, specialized test structures were fabricated in the university cleanroom. For this, metal–oxide–semiconductor (MOS) capacitors were used for reliability testing, because they represent a simplified structure of the MOS transistor in Fig. 3.1 and are well suited for this purpose. Like depicted in Fig. 3.3a, this can be realized by thermally oxidizing a highly doped silicon (Si) wafer for a high-quality silicon dioxide (SiO_2) dielectric and adding single aluminum (Al) contact pads to the topside and a global Al back contact to the backside. Whereas the back of the wafer is contacted via the chuck contact of the wafer prober, the individual topside pads are contacted with probing needles. Further elaborated test structures include a remote contact pad on a thicker SiO_2 field oxide and a guard ring around the active device region, like shown in Fig. 3.3b. This respectively improves the overall robustness of the device and defines the active region more accurately, preventing potential edge effects.

For this purpose, the capacitors were fabricated on highly doped n^{++}-Si 4″-wafers. The capacitor devices have a square shaped active region with a standard side length of 100 µm, an ultrathin insulating thermally oxidized SiO_2 dielectric of 2.7–4.0 nm thickness, and evaporated Al contacts of 500 nm thickness. The thicker SiO_2 dielectric of the elaborated test structure in Fig. 3.3b was also thermally grown and has a thickness of 500 nm. To implant the p^{++} guard ring well of 20 nm width, an initial n^{+}-Si wafer was used that was later additionally doped on the back side for a better electrical contact to the aluminum layer. The guard ring circumscribes the capacitor, not cropping the area of the active region. Further information can be found in the respective process plans of the devices in the appendix section A.3. Microscope images of these devices are shown in Figs. 3.3c and 3.3d.

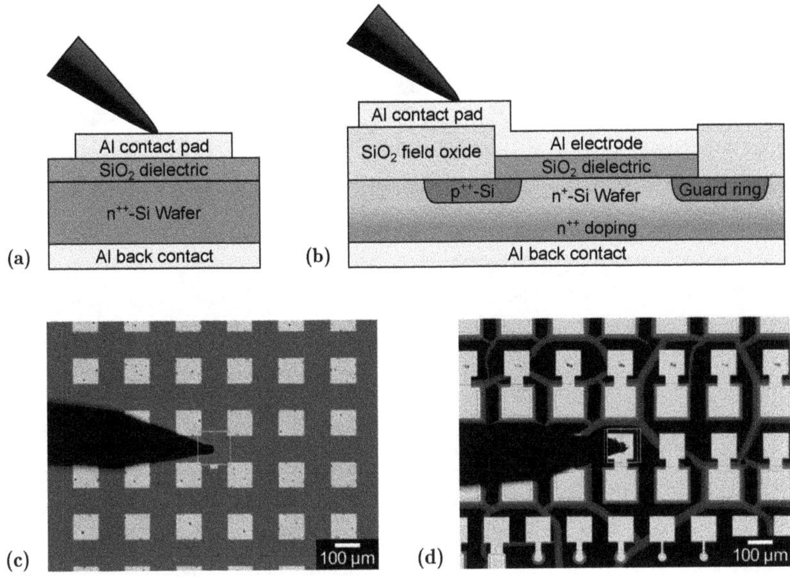

Figure 3.3: **(a)** Simple MOS capacitor test structure for TDDB testing. **(b)** Elaborate testing structure featuring a remote contact pad on a thicker dielectric and a guard ring to prevent edge-effects. The respective microscope images of **(c)** the simple (not contacted) and **(d)** the elaborate structure (contacted) depict the aluminum contact pads as light gray squares.

Probing over active area with a needle probe requires a systematic and consistent approach to not impact the reliability of the tested device. This is achieved by a so-called needle *overdrive*, depicted in Fig. 3.4a, which ensures a constant mechanical pressure of the needle on the contact pad. If the needle contact is too soft, a low-resistance electrical contact is not guaranteed, whereas a too hard contact could damage the device. Inconsistent mechanical pressure by the needle probe also impacts the lifetime of the device, as can be seen in Fig. 3.4b. A mechanical malfunction led to a deviating height setting during one measurement, resulting in affected reliability data (see Fig. 2.18b). To avoid such external influences on the device reliability, MOS capacitors with remote contact pads should be used.

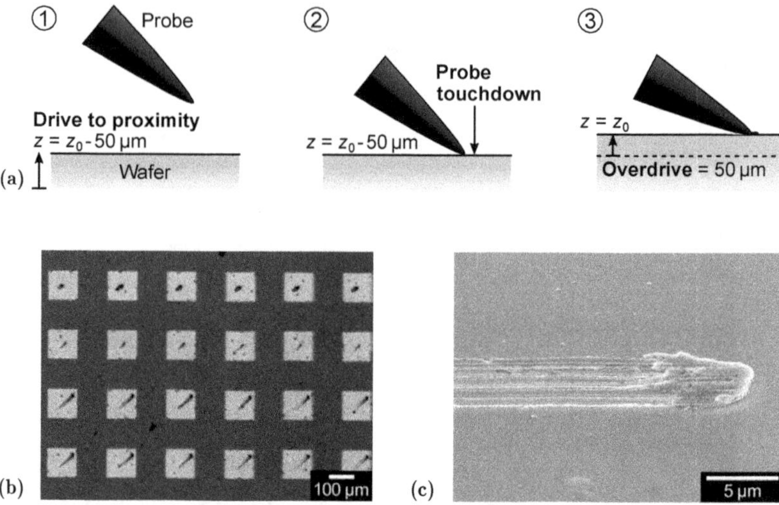

Figure 3.4: **(a)** Step sequence to achieve a constant needle overdrive of 50 µm for an operating height z_0 of the chuck. **(b)** Microscopy image of different needle scratch marks in the Al contact pads due to a mechanical malfunction between the upper and the lower half of the image. **(c)** Scanning electron microscopy image of a needle scratch in the Al contact pad.

As illustrated in Fig. 3.5, the TDDB experiments were conducted inside a FormFactor Summit 12000 needle-probe station (with MicroChamber® and AttoGuard® shielding technologies from electromagnetic interference and without illumination) and a thermal chuck ATT C60, controllable from −60 °C to +300 °C. For the measurements, the semiconductor device parameter analyzer Keysight B1500A is used in constant voltage mode and is equipped with B1517A high resolution source/monitor units (SMUs) with current measurement resolution of lower than 1 fA and voltage measurement resolution down to 0.5 µV, while supplying up to ±100 V and ±100 mA [Agi13]. For the rapid local temperature changes required in chapter 4, a Weller WTHA 1 hot air station is utilized, which has an electronically controlled, nominal temperature range of 50–600 °C.

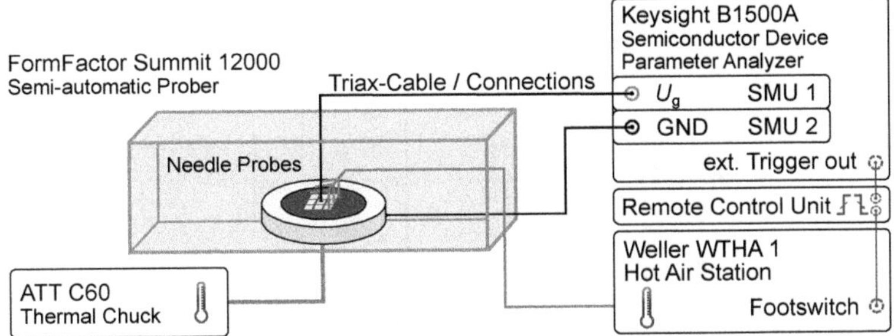

Figure 3.5: Experimental setup used for this work. The test wafer is electrically contacted via needle probes and the chuck bias. The device temperature can be controlled via the thermal chuck and a separate hot air station for rapid temperature changes. The reliability measurements are performed by a semiconductor parameter analyzer.

The loop resistance of the measurement setup is about 1–2 Ω, which leads to an RC time of < 1 ns for the used capacitors of about 0.1 nF. This is at least 3 orders of magnitude less than the shortest integration time of the SMUs, rendering the charging process of the capacitors insignificant for the measurements even in the case of large voltage steps. Attention must be paid to not change the impedance or capacitance of the measurement equipment or the the current compliance value for TDDB measurements, as this can influence the resulting lifetimes and failure distributions [Jac97].

In general, TDDB measurements can be performed in either accumulation or inversion, especially for constant voltage stress (CVS) tests [JED03]. However for ramp voltage stress (RVS), accumulation is recommended in order to minimize inversion capacitance effects [JED01]. For that reason, positive voltages are applied to the topside contact pad while grounding the backside contact via the chuck, resulting in accumulation of the MOS-capacity. This is depicted in Fig. 3.6 in a band diagram of the biased DUTs with applied gate voltage U_g larger than the flatband voltage U_{fb}, the Fermi energy E_f of the respective materials, the conduction band energy E_c, the valence band energy E_v, and band bending Ψ_s of the semiconductor Si, the applied electric field E_{ox} and potential difference U_{ox} for an electrical charge q across the SiO_2 layer. Only for thermal equilibrium of the Schottky barrier and negligible voltage drop across the semiconductor does U_{ox} equal U_g. E_{ox} can be derived from V_{ox} and the oxide thickness and must be corrected for U_{fb} and Ψ_s [JED03]. Furthermore, the band diagram is illustrated as an ideal condition without oxide or interface charges, mobile ions, or preexisting oxide trap states.

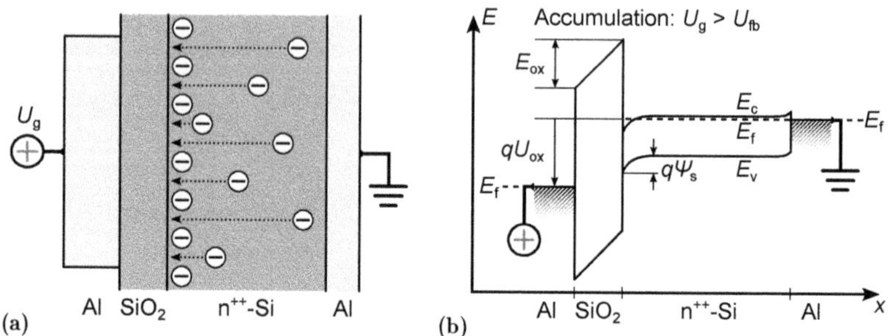

Figure 3.6: **(a)** Schematic material composition and **(b)** respective band diagram of the biased MOS capacitors in accumulation.

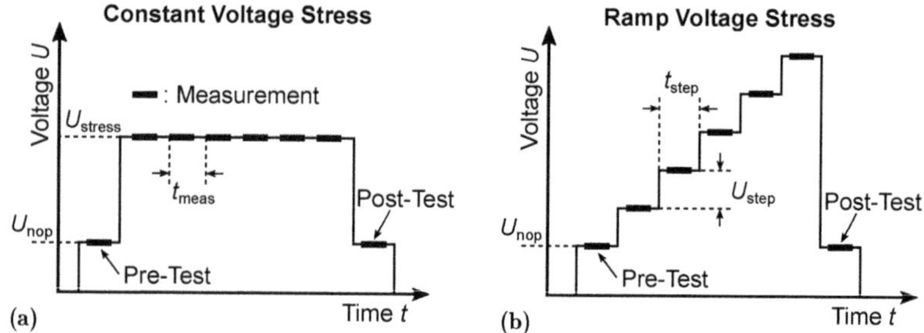

Figure 3.7: Measurement procedure for **(a)** CVS and **(b)** RVS as described in [JED03] and [JED01], respectively. The main stress measurement is bordered by integrity tests of the dielectric.

Before and after the main CVS or RVS measurement, the state of the device should be determined. As illustrated in Fig. 3.7, first an initial oxide integrity test can be conducted by applying normal operating voltage U_{nop} and checking if the measured current is within the operating specifications. After the main stress measurement, this integrity check can be repeated to verify an observed breakdown by its short-circuit behavior. These measurements can also be substituted by a careful I–U sweep like in Fig. 3.9.

Preceding a TDDB measurement series or full reliability characterization, a pre-characterization test [JED03] is conducted to find suitable stress voltages and currents. An exemplary RVS test is performed with the initial device type in Fig. 3.3a with a 3.7 nm thick oxide and shows three distinct regions in Fig. 3.8. For low voltages, usually a low constant plateau of leakage current is measured or if the leakage current is to small compared to the measurement resolution, only measure-

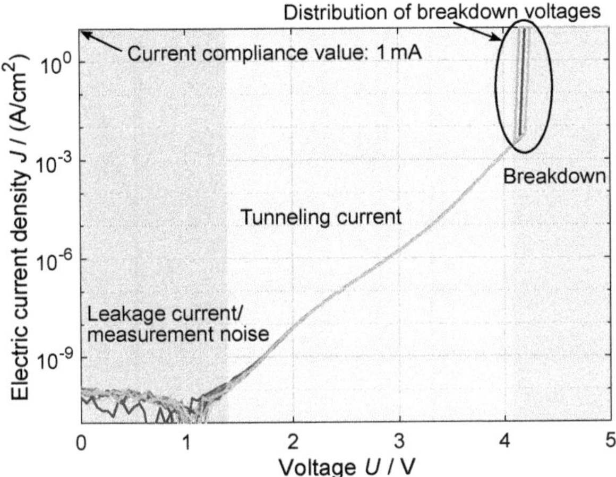

Figure 3.8: Pre-stress I–U characterization measurement of 13 DUTs. The curve is characterized by three different regions which differ in the behavior of the electric current density as indicated in the graph.

ment noise is recorded. At higher voltages, the I–U curve is dominated by tunneling current due to the thin oxide and applied voltage. This is due to direct tunneling, Fowler-Nordheim tunneling, thermionic emission, and for oxides of poor quality also by Frenkel-Poole emission [Sze06]. For even higher voltages, a hard dielectric breakdown is observed for all measured DUTs. As this was just a preliminary voltage sweep with large steps, the breakdown data are heavily grouped. A good start voltage for subsequent CVS testing is in the tunneling regime close to the observed breakdown voltages.

A representative TDDB measurement of the same wafer with CVS as well as the typical MOS capacitor I–U curve before and the short-circuit behavior of the DUTs after breakdown are depicted in Figure 3.9. The CVS measurement was conducted with a sampling rate of 50 Hz. The measurement points adjacent to the hard breakdown event show a current jump up to the current compliance value, exceeding the one order of magnitude criterion of a hard breakdown by far.

In conclusion, a sufficient basis of test structures and measurement methodology has been provided, which is used throughout this work to obtain the data for all shown Weibull distributions.

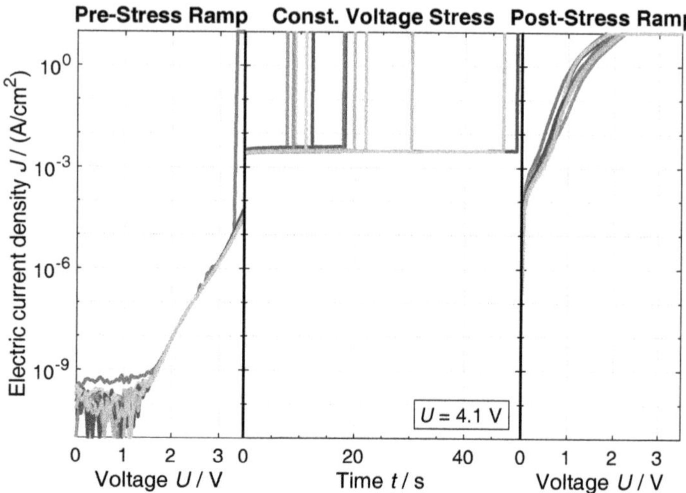

Figure 3.9: Reliability measurement of 13 DUTs for the failure mechanism TDDB. First, the pre-stress voltage ramp from 0–3.5 V shows characteristic *I*–*U* curves of MOS capacitors with an early breakdown of one device at 3.3 V that already exceeded the leakage current of all other measurements. Second, the CVS measurement at 4.1 V is conducted. The remaining DUTs exhibit a hard breakdown at Weibull distributed stress times. The final post-stress *I*–*U* curve verifies the breakdown of all devices, as a short-circuit behavior is observed.

3.3 Failure Acceleration

In this section, the different types of acceleration models necessary for the lifetime prediction as illustrated in Fig. 2.14 will be examined and specified for the TDDB failure mechanism. The procedure and measurements described here are conducted in analogy to those used for an industry technology qualification. Hence, the example data obtained in the university laboratories for in-house fabricated semiconductor devices are matched with industry TDDB reliability data for comparison, which have been generously made available by GLOBALFOUNDRIES. This demonstrates the good comparability of the applied industry-oriented processes and measurements in this work.

This reliability assessment is exemplarily conducted and presented on the simple MOS capacitor test structures as displayed in Fig. 3.3a with a SiO_2 thickness of 2.7 nm but also replicated for all other test structures used in this work. The presented test devices of the GLOBALFOUNDRIES 22FDX® fully depleted silicon on insulator (FD-SOI) technology with high-k metal-gate stacks on 12″-wafers are not mere MOS-capacity structures but complete MOS transistors as illustrated in Fig. 3.1. Furthermore, these devices are usually biased into inversion and form an inversion layer in the bulk material, with the result that a constant supply of charge carriers for the inversion layer is ensured via grounded source and drain contacts and no bulk contact is required [Str09]. For intellectual property reasons, explicit failure times and corresponding stress levels are not disclosed for GLOBALFOUNDRIES reliability data.

To begin with the reliability assessment of the university MOS devices, the available structures and performed stress measurements are listed in Table 3.1

Table 3.1: Stress matrix for MOS capacitors with 2.7 nm thick SiO_2 oxide used for the reliability assessment of section 3.3.

Voltage U	Temperature T	Area A
Voltage acceleration		
2.4 V		
2.5 V		
2.6 V	24 °C	$(100\,\mu m)^2$
2.7 V		
2.8 V		
Temperature acceleration		
	100 °C	
	125 °C	
2.1 V	150 °C	$(100\,\mu m)^2$
	175 °C	
Area scaling		
		$(70\,\mu m)^2$
		$(90\,\mu m)^2$
		$(100\,\mu m)^2$
		$(120\,\mu m)^2$
2.6 V	24 °C	$(140\,\mu m)^2$
		$(170\,\mu m)^2$
		$(190\,\mu m)^2$
		$(220\,\mu m)^2$
		$(240\,\mu m)^2$

The acceleration models discussed in this section and the given literature parameter values are all adopted from three major works by JEDEC [JED16], J. McPherson [McP12, McP19], and A. Strong [Str09] without claiming completeness.

Voltage Acceleration

The obtained Weibull distributions for CVS TDDB measurements with different voltages U_{stress} from Table 3.1 are depicted in Fig. 3.10 as a function of breakdown time t.

It is apparent that a higher voltage level accelerates the failure mechanism increasingly and reduces the expected lifetime. However, the question whether the voltage U_{ox} is the real stressor of the TDDB or just a virtual one and instead the electrical field E_{ox} is the real stressor is not definitely settled and depends on the assumed acceleration model. The origin of these models is either an empirical one, which is later justified by a physical foundation, or started out from a physical consideration that lead to a mathematical formulation.

The *exponential E model* or also called *thermochemical E model* states that field enhanced thermal bond-breakage inside a dielectric of $> 4\,\mathrm{nm}$ is the cause of TDDB, especially for low electric fields $< 10\,\mathrm{MV/cm}$, and is believed to be the main mechanism for long-term degradation under use conditions. The time-to-failure (TTF) is obtained by

$$\mathrm{TTF} = A_0 \cdot \exp\left(-\gamma(T) \cdot E_{ox}\right) \tag{3.1}$$

with an arbitrary scale factor A_0, a temperature dependent field acceleration parameter $\gamma(T) \approx 2.5$–$3.5\,\mathrm{cm/MV}$, and the applied field E_{ox} across the dielectric.

The *exponential $\sqrt{E}$ model* assumes a current-induced dielectric degradation that is caused by Poole-Frenkel or Schottky emission and is often associated with silica oxides of poorer quality that have preexisting trap states in the oxide which facilitate tunneling. The model is formulated as

$$\mathrm{TTF} = D_0(T) \cdot \exp\left(-\alpha \cdot \sqrt{E_{ox}}\right) \tag{3.2}$$

with an arbitrary temperature dependent scale factor D_0, a root-field acceleration parameter $\alpha(T)$, and the applied field E_{ox} across the dielectric. The respective low-k dielectrics are often used in the BEOL and the corresponding testing methodology is given in [JED10].

The *power-law U model* or so-called *anode hydrogen release model* presumes bond breakage and release of hydrogen ions at the Si–SiO_2 interface causing them to drift away towards to cathode and to create defects in the dielectric. This model was originally proposed for ultrathin dielectrics $< 4\,\mathrm{nm}$ but has been extended for thicknesses up to $10\,\mathrm{nm}$. The case is made that ballistic transport of electrons to the anode delivers the energy needed for this bond breakage and is therefore dependent on the voltage as

$$\mathrm{TTF} = B_0(T) \cdot (U_{ox})^{-n} \tag{3.3}$$

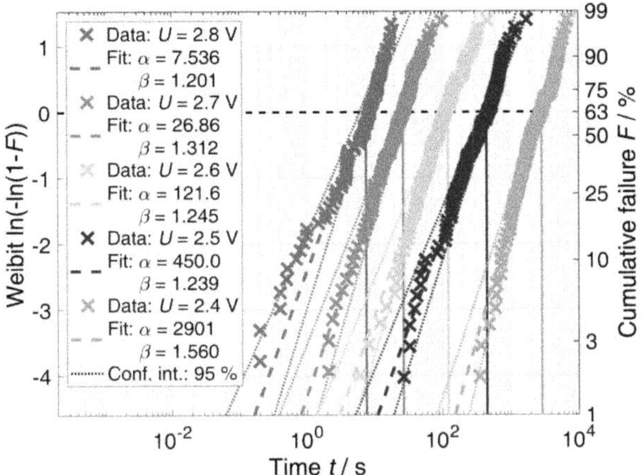

(a) Failure times of the MOS capacitor test devices.

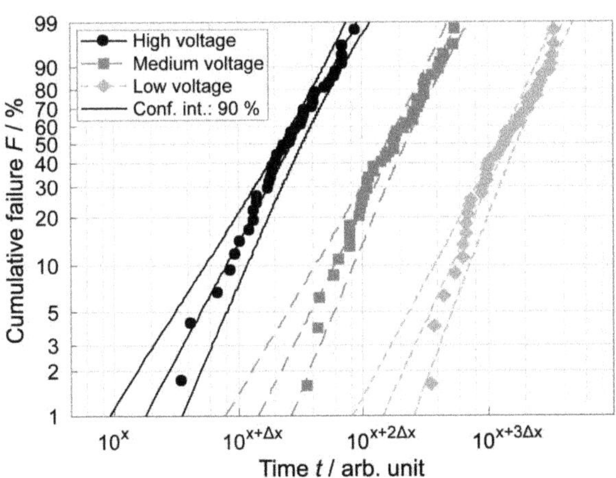

(b) Failure times from GLOBALFOUNDRIES 22FDX® technology qualification.

Figure 3.10: Weibull distributed TDDB failure data from DUTs for different stress voltages with corresponding confidence intervals.

with an arbitrary temperature dependent scale factor $B_0(T)$, the applied voltage U_{ox} across the dielectric, and the voltage acceleration exponent $n \approx 40$–48, which is observed to decrease for thicker dielectrics.

Finally, the *exponential* $1/E$ *model*, also known as *anode hole injection model*, is the most popular current-induction model, postulating that for direct and Fowler-Nordheim tunneling in high-quality SiO_2 electrons from the cathode cause degradation in the dielectric by impact ionization and additionally produce holes at the anode which tunnel back to the cathode creating further defects. The lifetime is given as

$$\mathrm{TTF} = \tau_0(T) \cdot \exp\left(\frac{G(T)}{E_{ox}}\right) \tag{3.4}$$

with an arbitrary temperature dependent scale factor $\tau_0(T)$, an temperature dependent inverse field acceleration parameter $G(T) \approx 300$–$350\,\mathrm{MV/cm}$, and the applied field E_{ox} across the dielectric.

All these models are differently well justified by a physical framework and measurements, exhibit different weaknesses in explaining certain experimental observations and are validated over different, partially overlapping stress levels and oxide thickness ranges, and are sometimes found to better fit voltage than electric field dependencies for thin dielectric layers [Str09, Wu09a, Wu09b, McP12, JED16, McP19]. Furthermore in an attempt to unify the TDDB models and to explain all experimental observations, complementary and unifying models have been proposed combining field-induced and current-induced models to fit reliability data over all experimentally observable time frames [Hu99, McP00, Cro15, Oka17].

In order to evaluate the stress acceleration measured in Fig. 3.10 and find an adequate voltage model, the obtained $\boldsymbol{t_{63}}$ data can be fitted with a linearized formula of the presented equations. To facilitate this, the arbitrary scale factors A_0, D_0, B_0, and τ_0 are substituted by the scale factor a, and the stress acceleration parameters γ, α, n, and G are replaced by b, resulting in the linearized acceleration models fitted in Fig. 3.11 and one carefully selected model in Fig. 3.12:

$$\text{Exponential } E \text{ model:} \quad \ln(\mathrm{TTF}) = (-b \cdot E) + \ln(a) \tag{3.5a}$$

$$\text{Exponential } \sqrt{E} \text{ model:} \quad \ln(\mathrm{TTF}) = \left(-b \cdot \sqrt{E}\right) + \ln(a) \tag{3.5b}$$

$$\text{Power-law } U \text{ model:} \quad \ln(\mathrm{TTF}) = (-b) \cdot \ln(U) + \ln(a) \tag{3.5c}$$

$$\text{Exponential } 1/E \text{ model:} \quad \ln(\mathrm{TTF}) = (b/E) + \ln(a) \tag{3.5d}$$

The estimated model parameters $\hat{a}$ and $\hat{b}$ are given in Table 3.2 and all four voltage acceleration models are combined in Fig. 3.13 for TTF extrapolation comparison.

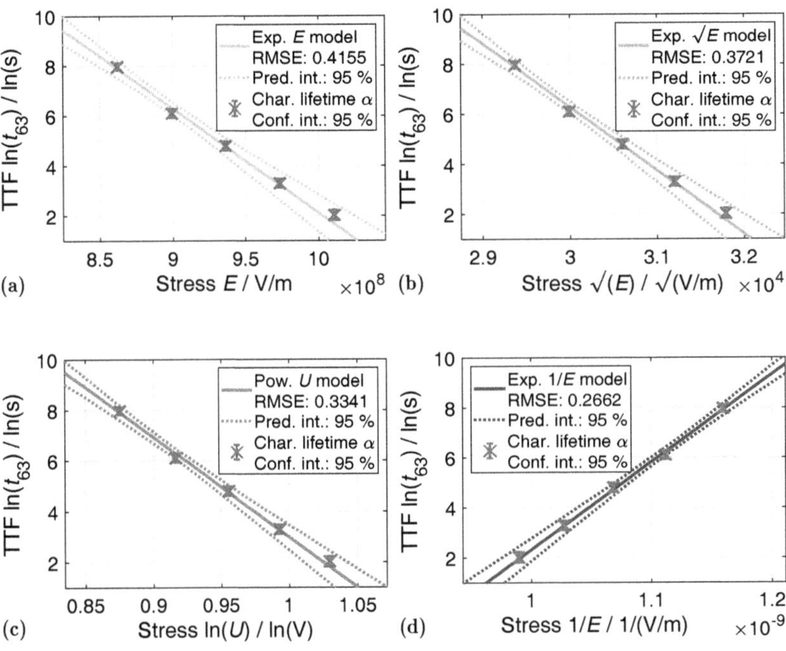

Figure 3.11: Characteristic lifetimes t_{63} from Fig. 3.10a fitted with different linearized voltage acceleration models: **(a)** Exponential E model, see Eq. (3.5a). **(b)** Exponential $\sqrt{E}$ model, see Eq. (3.5b). **(c)** Power-law U model, see Eq. (3.5c). **(d)** Exponential $1/E$ model, see Eq. (3.5d).

Table 3.2: Voltage acceleration model fit parameters given in their respective units with the dimensions s, V, and m.

Model:	Exp. E fit	Exp. $\sqrt{E}$ fit	Pow. U fit	Exp. $1/E$ fit
CVS analysis and parameter fitting (see Fig. 3.13)				
$\hat{a}$	$\exp(44 \pm 6)$	$\exp(83 \pm 11)$	$\exp(43 \pm 5)$	$\exp(-33 \mp 4)$
$\hat{b}$	$(4.2 \pm 0.7) \cdot 10^{-8}$	$(2.5 \pm 0.4) \cdot 10^{-3}$	$(4.0 \pm 0.5) \cdot 10^{1}$	$(3.5 \pm 0.4) \cdot 10^{10}$
Literature values [Str09, JED16, McP19]				
$\hat{b}$	$(2.5–3.5) \cdot 10^{-8}$	—	$(4.0–4.8) \cdot 10^{1}$	$(3.0–3.5) \cdot 10^{10}$

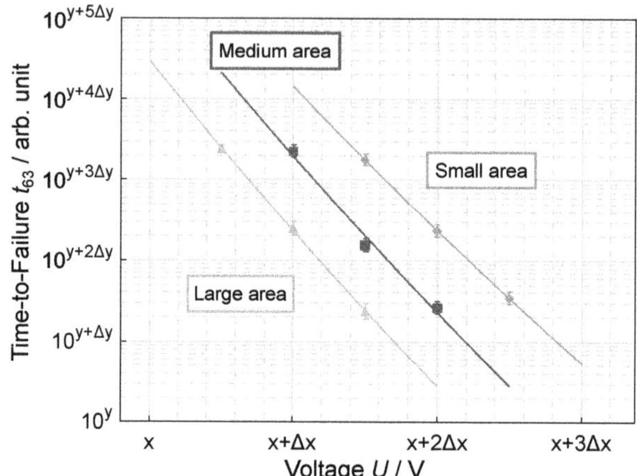

Figure 3.12: Characteristic lifetimes t_{63} from Fig. 3.10b and other devices with different gate areas simultaneously fitted with a selected voltage acceleration model by GLOBALFOUNDRIES.

It can be seen that the above models in Eqq. (3.1)–(3.4) rank in ascending order from conservative extrapolations to very optimistic long extrapolated lifetimes. In addition, the prediction intervals for the two plots differ. In Fig. 3.13a, functional observation bounds are given, indicating the uncertainty for the next function value $f(x_{n+1})$, while in Fig. 3.13b, observational prediction bounds are given. The observation bounds are wider than the functional bounds because they give the uncertainties of the predicted fitted curve as well as an additional randomly distributed variation for a new observation $y_{n+1}(x_{n+1})$. For lifetime extrapolations, as intended in Fig. 2.14, the observational prediction bounds are chosen to be more favorable.

As far as the fit of the different models is concerned, it can be seen in Fig. 3.11 from the root mean squared error (RMSE) values, which are used here as GOF figure of merit, that the more optimistic models, especially the $1/E$ model, fit best. Regardless, it remains that all fitted models fit relatively well in the fit domain, only extrapolating to lower stress and long lifetimes yields large differences. This has to be taken into account when choosing a suitable model.

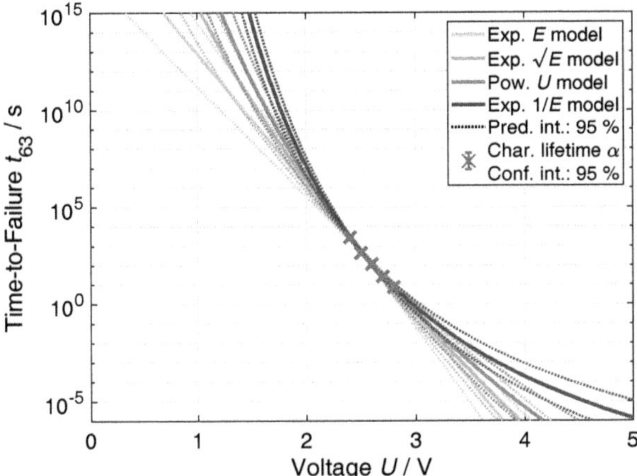

(a) Voltage acceleration models extrapolated with functional prediction intervals.

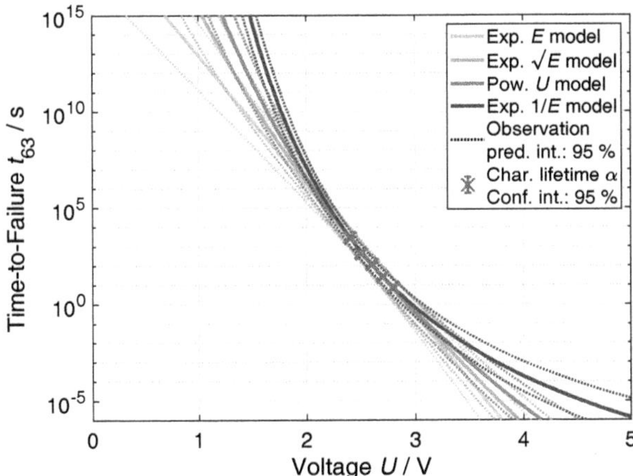

(b) Voltage acceleration models extrapolated with observation prediction intervals.

Figure 3.13: Time-to-failure extrapolation plots with all four voltage acceleration models combined.

Temperature Acceleration

The other important stressor for TDDB is the temperature. The conducted temperature measurements in Fig. 3.14 with identical CVS at four different temperatures reveal that temperature stress increasingly accelerates the TDDB degradation and shortens the lifetime of the DUTs.

In contrast to the variety of voltage and electric field acceleration models, by common consent the *Arrhenius model* fits the temperature acceleration sufficiently good, bearing the temperature dependent voltage acceleration parameters in mind, which may already include the Arrhenius relation, as

$$\mathrm{TTF} = a \cdot \exp\left(\frac{Q_{\mathrm{a}}}{k_{\mathrm{B}} T}\right) \tag{3.6}$$

with the arbitrary scale factor a, the activation energy Q_{a}, the Boltzmann constant $k_{\mathrm{B}} = 8.617386 \cdot 10^{-5}\,\mathrm{eV/K}$, and the device temperature T in Kelvin. Hence, Q_{a} represents the temperature acceleration parameter of the Arrhenius model and is often referred to as *apparent activation energy*. This is due to the fact that Q_{a} does usually not correspond to a single physically justified activation energy of a single thermally activated reaction but is a conglomeration of all concurrent reactions and degradation mechanisms. Hence, the observed apparent activation energy can change with electric field or temperature. Nevertheless, it is usually obtained as a fit parameter of around 0.6–0.9 eV for TDDB in SiO_2 dielectrics.

The Arrhenius model has been fitted to the measures $t_{\mathbf{63}}$ failure times in Fig. 3.15 as linearized relation

$$\ln(\mathrm{TTF}) = \frac{Q_{\mathrm{a}}}{k_{\mathrm{B}} T} + \ln(a) \tag{3.7}$$

and results in the model parameters $a = 1.0 \cdot 10^{-08}\,\mathrm{s}$ and $Q_{\mathrm{a}} = 0.79\,\mathrm{eV}$.

The TTF–temperature relation is plotted in Fig. 3.16 with the already introduced observation prediction intervals and can be used for lifetime extrapolation to use conditions.

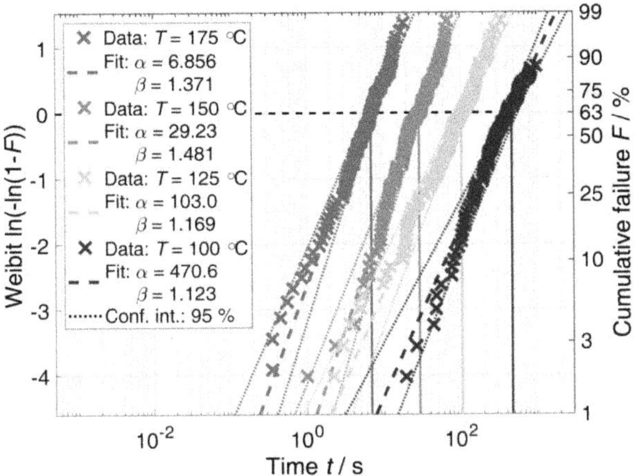

(a) Failure times of the MOS capacitor test devices.

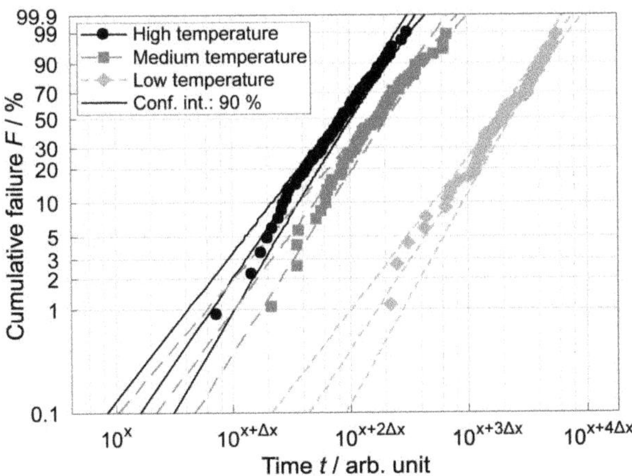

(b) Failure times from GLOBALFOUNDRIES 22FDX® technology qualification.

Figure 3.14: Weibull distributed TDDB failure data from DUTs for different stress temperatures with corresponding confidence intervals.

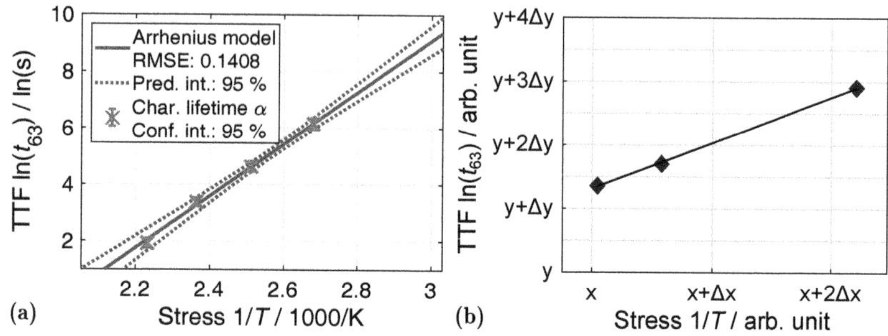

Figure 3.15: Characteristic lifetimes t_{63} from Fig. 3.14 fitted with the linearized Arrhenius model in Eq. (3.7): **(a)** Fit of the MOS capacitor data in Fig 3.14a. **(b)** GLOBALFOUNDRIES fit of the 22FDX® data in Fig. 3.14b.

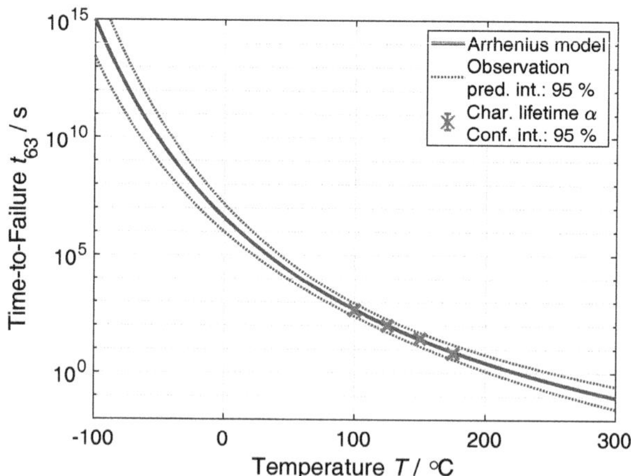

Figure 3.16: Time-to-failure extrapolation plot with the Arrhenius temperature acceleration model and observation prediction intervals.

Area Scaling

After determining the two stress acceleration models, another quantity that influences TDDB lifetime is the dielectric area of the device. Since test structures are usually smaller in area than whole integrated circuits, this must be considered in any reliability qualification.

The *Poisson* area scaling model is universally accepted for the failure mechanism TDDB, as it describes the relation of different areas with randomly distributed defects to device lifetime. This assumption of uniformly distributed defects holds for intrinsic and usually also for extrinsic defects of TDDB. However for clustered defects, other area scaling models like the Seeds or Stapper model might be adequate.

According to the Poisson model, the resulting cumulative failure percentage F for a different device area is derived as

$$F = 1 - (1 - F_{\text{ref}})^{\frac{A}{A_{\text{ref}}}} \tag{3.8}$$

with the cumulative failure percentage F_{ref} and the area of the reference device under identical stresses A_{ref}, and the area of the new structure A.

This means that plotted CDFs for different device areas are shifted vertically in the probability plot. In particular, the Poisson model preserves the distribution's shape and is therefore independent of the failure distribution. Measurements in Fig. 3.17 for different device areas are shifted on the area-normalized Weibull axis accordingly to the Poisson scaling and result in a joint EDF over a large Weibit range, which can be used for a better estimation of the Weibull slope β.

Inserting the area scaling of Eq. (3.8) into the equation of the Weibull CDF (see Eq. (2.6b)), an expression for the lifetime can be derived as

$$\text{TTF} = t_{\text{ref}} \cdot \left(\frac{A}{A_{\text{ref}}}\right)^{-\frac{1}{\beta}} \tag{3.9}$$

with the lifetime of the reference device t_{ref}, and the Weibull slope β, which can also be directly applied for lifetime extrapolation.

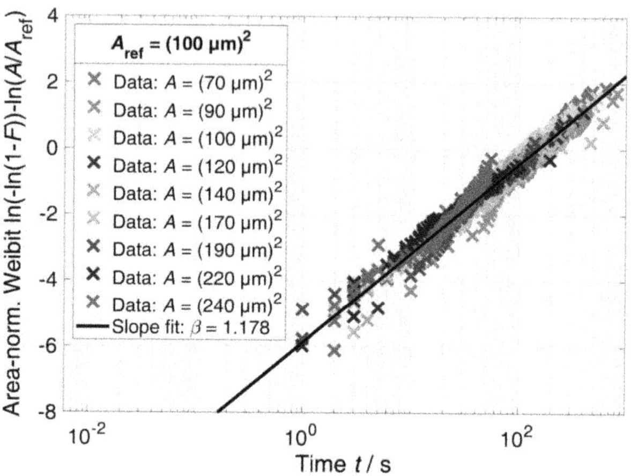

(a) Failure times of the MOS capacitor test devices.

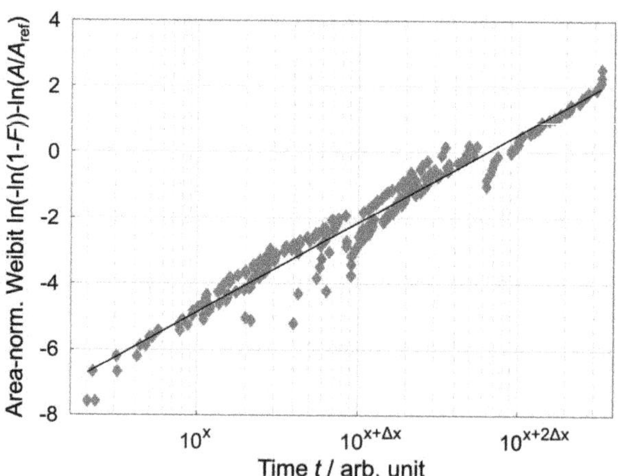

(b) Failure times from GLOBALFOUNDRIES 22FDX® technology qualification.

Figure 3.17: Area scaling of Weibull distributed TDDB failure data of DUTs with different gate areas. The Weibit axis is area normalized according to the Poisson model. These plots are often used to fit the Weibull slope β with more data and over a wider Weibit range.

Resulting Acceleration Factors and Lifetime Projection

Concluding this section, the different acceleration models can now be combined for the final lifetime projection of TDDB for a semiconductor component under operating conditions. Therefore, the applied acceleration models are used to calculate the respective acceleration factors (see Eq. (2.5)) for voltage AF_U, temperature AF_T, area scaling AF_A, and Weibull scaling AF_F.

For illustration, the measured failure distribution for $T_{\mathrm{acc}} = 125\,°\mathrm{C}$ and $U_{\mathrm{acc}} = 2.1\,\mathrm{V}$ is used as a reference measurement to extrapolate to arbitrary operating conditions. We assume a Weibull distribution with the parameters $t_{\mathbf{63},\mathrm{acc}} = 103.0\,\mathrm{s}$ and $\beta = 1.178$ and the exponential $1/E$ model with $b = 3.5\cdot 10^{10}$ for voltage acceleration. An activation energy of $Q_{\mathrm{a}} = 0.79\,\mathrm{eV}$ is employed for the Arrhenius model and the reference area is $A_{\mathrm{ref}} = (100\,\mu\mathrm{m})^2$. The arbitrarily chosen use conditions are $T_{\mathrm{nop}} = 80\,°\mathrm{C}$ and $U_{\mathrm{nop}} = 1.2\,\mathrm{V}$ for a chip area $A_{\mathrm{chip}} = 1\,\mathrm{cm}^2$. The failure requirement is set to $F = 10^{-6} = 1\,\mathrm{ppm}$.

The resulting total acceleration of the failure mechanism and the associated lifetime prediction are calculated as follows:

$$\mathrm{TTF} = t_{\mathbf{63},\mathrm{acc}} \cdot \mathrm{AF}_U \cdot \mathrm{AF}_T \cdot \mathrm{AF}_A \cdot \mathrm{AF}_F \tag{3.10}$$

$$= t_{\mathbf{63},\mathrm{acc}} \cdot \exp\left[b\left(\frac{1}{E_{\mathrm{nop}}} - \frac{1}{E_{\mathrm{acc}}}\right)\right] \cdot \exp\left[\frac{Q_{\mathrm{a}}}{k_{\mathrm{B}}}\left(\frac{1}{T_{\mathrm{nop}}} - \frac{1}{T_{\mathrm{acc}}}\right)\right] \cdot \left(\frac{A_{\mathrm{chip}}}{A_{\mathrm{ref}}}\right)^{-\frac{1}{\beta}} \cdot \left[-\ln\left(1-F\right)\right]^{\frac{1}{\beta}} \tag{3.11}$$

$$= 103.0\,\mathrm{s} \cdot 4.544\cdot 10^{14} \cdot 1.880\cdot 10^{1} \cdot 4.022\cdot 10^{-4} \cdot 8.065\cdot 10^{-6}$$

$$= 2.854\cdot 10^{9}\,\mathrm{s}$$

The expected lifetime of these devices under the given use conditions is derived as $\mathrm{TTF} = 2.854\cdot 10^9\,\mathrm{s} \approx 92.8\,\mathrm{years}$ and would subsequently be compared with a required product lifetime and with respect to corresponding safety margins.

Additionally, this lifetime projection can be visualized in accordance to Fig. 2.14, like illustrated in Fig. 3.18 in which all measured Weibull distributions are plotted and extrapolated to the respective stress levels and device areas of the final product under use conditions. As a next step, the overall fitted Weibull slope is extrapolated to lower failure fractions, where it is required to not intersect with the requirement targets of the reliability qualification.

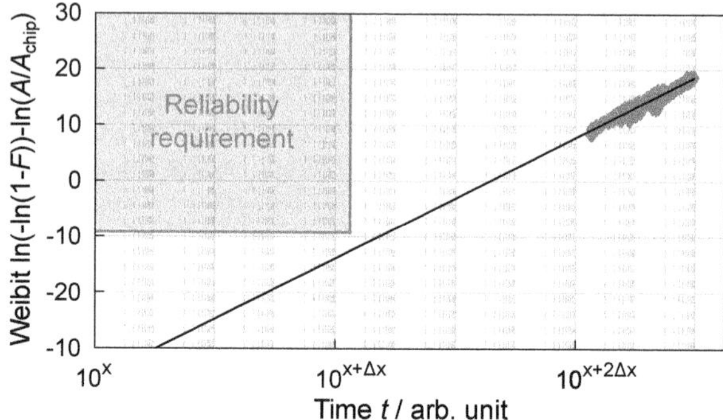

Figure 3.18: End-of-life projection by GLOBALFOUNDRIES with all available TDDB qualification measurements extrapolated to the maximal specified operating conditions. The reliability requirement target for the qualification, which consists of an expected lifetime for a certain failure quantile, is illustrated as red box that has to be exceeded. Consequently, the technology qualification target has been achieved.

The procedure shown here corresponds to the standardized status quo of the reliability qualification which is accepted and common practice for TDDB according to the AEC-Q100 [Aut14]. However, this method is based on the assumption of constant stress as reliability requirement under use conditions. Since not all application specific environmental and functional stressors can sufficiently be represented and taken into account in this way, more and more emphasis is being placed on application profile specific qualification. This aspect will be addressed and discussed in detail in the following chapter.

4

Effective Stressors

4.1 Mission Profiles

The automotive industry is steering into a future of new, innovative challenges with driver assistance systems, autonomous driving, electric mobility and car connectivity on their way into the market. In order to realize such ambitious new features, the usage of leading-edge semiconductor technologies is essential. To further on secure the typical high reliability standards requested in the automotive industry, great efforts are made to standardize and advance existing qualification processes. One example are the stress test qualification plans of the Automotive Electronics Council (AEC). In addition to standardized test conditions, they also consider customer mission profile based qualification and robustness validation in case the standard conditions in the AEC-Q100 are not sufficient to cover the application's lifetime requirements [Aut14].

Mission profiles are usually a conglomeration of all relevant stresses on the chosen product in form of time series, histograms or effective stressors (see Fig. 4.1). Due to the extended duty cycles, e.g. for electric mobility applications that are predicted for the future, standard test times may not be sufficient anymore and other ways of qualification must be considered. Either a thorough robustness validation [Kan14] is performed or the already hastened qualification test must be further accelerated. The latter is not always possible, because at higher stresses other usually insignificant failure mechanisms can become the dominant reason for failure or other components of the test device, for instance packaging at high temperatures, will fail due to enhanced stresses. Besides that, the degree of detail of mission profiles is constantly increasing to distinguish e.g. different installation locations or supplier adjustments.

While the use of mission profiles becomes increasingly necessary, as already stated, literature on the physical justification for the creation of histograms and effective qualification stresses is barely available.

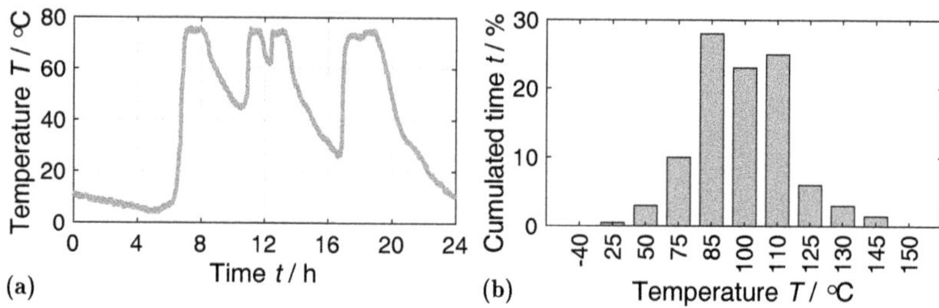

Figure 4.1: Different types of mission profiles: **(a)** An exemplary stress–time diagram and **(b)** an exemplary stress bar chart.

Even the recently added chapter on "Conversion of Dynamical Stresses into Effective Static Values" in [McP19] does not report any experimental data nor does it reference any literature on that subject.

Usually in mission profiles, a typical daily stress–time profile is repeated many times and finally cumulated in a stress histogram used as reliability requirement. Due to the fact that the stress in mission profiles cannot be assumed to be constant, cumulative damage models are needed for the correct evaluation of such mission profiles. Thus, a variety of so-called cumulative damage models have been proposed in literature.

The main aspect of publications on this topic is mostly of statistical nature. This includes topics like stress-test planning and optimization and Bayesian or maximum likelihood estimation (MLE) fitting methods [Tan96, Xio99, Kwe99, Li07, Ham15, Mil83]. Various cumulative damage models have also been discussed with focus on different stress acceleration models [Zha05]. However, only little attention is given to the experimental validation of those proposed models. Nelson, who developed the cumulative exposure model [Nel80], already mentioned the lack of validation nearly 30 years ago [Nel90] and it is still an important issue that requires ongoing attention [Pen10].

Therefore, three of the most accepted approaches of cumulative damage models, namely the cumulative exposure (CE) [Nel80], the tampered random variable (TRV) [DeG79], and the tampered failure rate (TFR) model [Bha89], will be introduced in analogous ways and subsequently compared to experimental findings. With the focus of empirical validation in mind, a suitable procedure is cyclical stress testing (CST). By applying this method of alternating stress levels, divergent failure behavior can be emphasized independently of failure mechanism and underlying acceleration model. Hence, the applicability of the aforementioned cumulative damage models for semiconductor reliability will be examined in theory and experiment on the example of time-dependent dielectric breakdown (TDDB) with the state-of-the-art 22 nm high-k metal-gate technology from GLOBALFOUNDRIES and university metal–oxide–semiconductor (MOS) capacitors. Furthermore, a concept of transforming mission profiles and alternating step-stress accelerated life testing (SSALT) stress sequences into equivalent effective stress levels or effective stress times for the two stressors of TDDB, voltage and temperature, is presented and investigated. This concept is then extended onto multi-dimensional mission profiles with combined stressors.

4.2 Cumulative Damage Models

As mentioned before, cumulative damage models were developed in various forms for different areas of application. The key difference between them is the type of transformation regarding each consecutive pair of stress steps. It may be a transformation of time [DeG79], an acceleration of failure rate [Bha89], or a connection condition of a different kind [Nel80]. Although the following cumulative damage models can be applied to miscellaneous failure distributions, only the Weibull distribution is employed in this work for the purpose of elaborating equations.

4.2.1 Cumulative Exposure Model

The cumulative exposure (CE) model is one of the basic and well-established cumulative damage models. It was initially developed and shown on data of cable insulation breakdown. The CE model states that the remaining life depends only on the current cumulative fraction failed and the cumulative distribution function (CDF) of the current stress. Therefore, the successive failure behavior is independent of how previous fails accumulated. [Nel80]

The resulting CDF F_{CE} is merged step-wise from the CDFs of each individual stress step F_i which corresponds to the duration between t_i and t_{i+1}

$$F_{\mathrm{CE}}(t) = \begin{cases} F_1(t), & t \leq t_2 \\ F_2\left(t - t_2 + t_2'\right), & t > t_2 \end{cases} \tag{4.1}$$

where t_2' denotes the equivalent starting time of F_2. The connecting condition is satisfied by $F_2(t_2') = F_1(t_2)$.

By introducing $\tau_i = t_i - t_i'$ as additional location parameter of the observed CDF, a simpler, recursive form of the CE model emerges:

$$F_{\mathrm{CE}}(t) = F_i\left(t - \tau_i\right), \qquad t_i < t \leq t_{i+1} \tag{4.2}$$

where the equivalent starting time t_i' can be obtained by

$$t_i' = (t_i - \tau_{i-1}) \cdot \mathrm{AF}_i \tag{4.3}$$

with τ_1 being the time offset at which the first stress is applied in respect to the start of the measurement (usually $\tau_1 = 0$) and the acceleration factor AF of the stress levels i and $i-1$, which is defined as the ratio of lifetimes of the respective CDFs that correspond to the same cumulative fractions failed (see Eq. (2.5)):

$$\mathrm{AF}_i = \frac{\alpha_i}{\alpha_{i-1}} \tag{4.4}$$

An equivalent recursive representation was also deduced in [Aal07] for the application on ramp voltage stresses.

When the failure distribution in question is a Weibull distribution, the newly obtained CDF can be interpreted as a three-parametric Weibull distribution with the formula

$$F_i(t) = 1 - \exp\left\{-\left(\frac{t-\tau_i}{\alpha_i}\right)^\beta\right\} \tag{4.5}$$

where α denotes the t_{63} time and β is the slope or shape parameter of the Weibull distribution. It is assumed that β is constant between different stress levels, i.e. the failure mechanism is not changed. This assumption also extends to the following stress models and is commonly presumed in literature on this topic, even if not explicitly stated.

While Nelson applied his model to Weibull distributions and an inverse power law stress–lifetime relationship [Nel80], the CE model can be generalized [Zha05] and applied to many other acceleration models.

Graphically, the CE model can be depicted as a horizontal shift of the new failure distribution F_2 towards the initial distribution F_1 in order to fulfill the connection condition at the time of stress change t_2 (see Fig. 4.2). The entire distribution is shifted by the same time τ_2, which is defined as the difference between the switching point t_2 and the equivalent starting time t_2' at the current stress. Due to the logarithmic time axis, the shifted distribution features a bent behavior.

To summarize, the most prominent feature of the CE model is that the current behavior of the specimen does only depend on the current stress, whereas the history of cumulative pre-stresses only matters as to what cumulative fraction the population has already failed. Hence, the CE model is commutative in regards to the stress sequence as the different stress distributions of consecutive stress levels are stitched together piecewise.

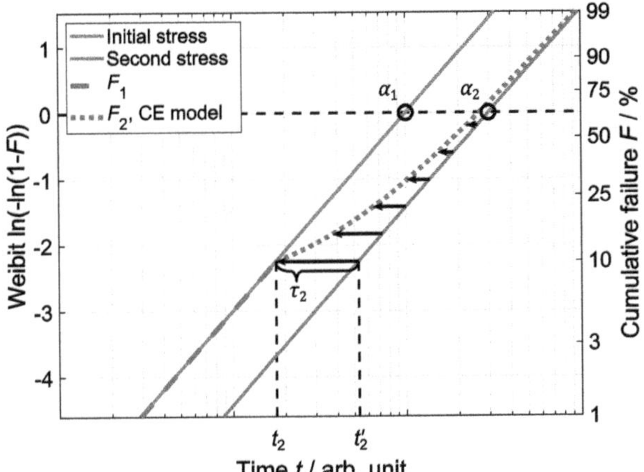

Figure 4.2: The CE model connects two consecutive stress level steps by shifting the succeeding stress distribution by an interval τ_i in time to intersect with the precessing stress level distribution at the time of stress change.

4.2.2 Tampered Random Variable Model

As the name tampered random variable (TRV) suggests, this model tampers with a variable, which in this case is the time t. In particular, the remaining life of a specimen after the stress change is multiplied with a tampering coefficient $\kappa_i > 0$ [DeG79]. Thereby, the remaining life is shortened by κ_i or prolonged by κ_i^{-1}, respectively:

$$t_{\mathrm{TRV}} = \begin{cases} t, & t \leq t_2 \\ t_2 + \kappa_2 \left(t - t_2\right), & t > t_2 \end{cases} \tag{4.6}$$

This time transformation can be directly applied to the Weibull CDF and results in:

$$F_{\mathrm{TRV}}(t) = \begin{cases} F_1(t), & t \leq t_2 \\ F_1 \left(t_2 + \kappa_2 \left(t - t_2\right)\right), & t > t_2 \end{cases} \tag{4.7}$$

Furthermore by definition, the tampering coefficient κ_i equals the inverse of the acceleration factor AF_i when applied to Weibull distributions of step-stresses:

$$\kappa_i = \frac{1}{\mathrm{AF}_i} \tag{4.8}$$

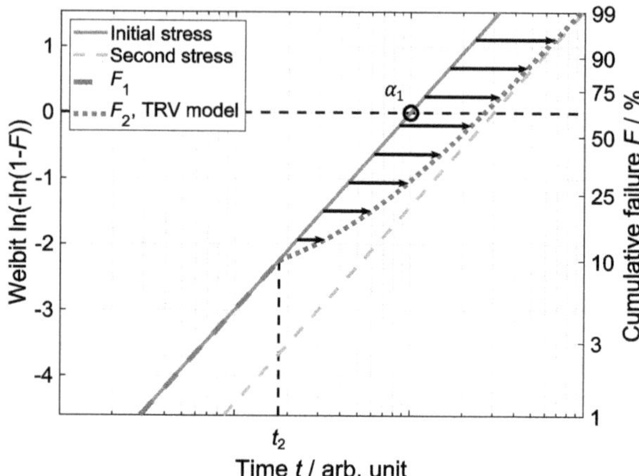

Figure 4.3: The TRV model shortens or prolongs the remaining lifetime of a specimen after a stress level change by transforming the time variable within the CDF of the initial stress distribution. The CDF of the subsequent distribution is involved only indirectly by means of determining the tampering coefficient κ_i and is therefore represented by a half-transparent dashed line.

In contrast to the formulation of the CE model in Eq. (4.2), the behavior of the TRV model can be solely described by using the Weibull parameters of the first stress distribution. This appears to be a more transparent physical motivation for cumulative damage behavior than the one offered by the CE model [Bha89]. Aside from that, both models can be motivated likewise by shock models and wear-out processes [Sha83], although the TRV model was initially not supported by experimental data.

As later described in the section 4.2.5 on model comparison, the TRV and CE model coincide for our utilization. Thus, an explicit formula for multiple step-stresses does not have to be developed for the purpose of this work, as the model behavior in the later chapters is calculated numerically.

Unlike the CE model, the TRV model uses the scale parameter of the subsequent stress only for calculating the tampering coefficient κ_i. As depicted in Fig. 4.3, starting from the initial failure distribution F_1, the behavior of this distribution is then modified after the stress change by stretching or compressing the initial distribution on the time axis with factor κ_2.

In conclusion, although differing in physical motivation and mathematical formulation at first, the time transformation of the TRV model exhibits the same results as the CE model that also approaches the distribution of the second stress level asymptotically on a logarithmic time axis. This of course only holds true for the assumption that the failure mechanism does not change for different stress levels.

4.2.3 Tampered Failure Rate Model

In distinction to the previously described cumulative damage models, the key point of the tampered failure rate (TFR) model is the focus on the hazard function or failure rate $h(t)$ (see Eq. (2.3)) of the distribution. This approach to cumulative damage with focus on failure rate in contrast to failure time correlates with the principle of various physical failure acceleration models [Sha14]. It is assumed that by applying a different stress level, the failure rate is changed by a factor $k_i > 0$:

$$h_{\text{TFR}}(t) = \begin{cases} h_1(t), & t \leq t_2 \\ k_2 \cdot h_1(t), & t > t_2 \end{cases} \tag{4.9}$$

The tampering coefficient k_i depends on the concerned stress levels and possibly on the time of stress change t_i, which is not further elaborated by the authors. Although the idea of tampering with the hazard function at stress level change was most famously publicized in Cox's proportional hazard model [Cox72], it has not been parameterized for use and analysis of step-stress testing prior to the TFR model. However, the TFR model was initially only considered in the context of a two-step partially accelerated life test without any validation by experimental data, so an iterative function for a multiple step-stress setting was not specified. [Bha89]

Nevertheless, because such a function is needed for the purpose of modeling the CDF of multiple or alternating step-stress levels, a general term is developed for the TFR model. When applied to Weibull distributions, the hazard function h_i (see Eq. (2.6c)) is defined as

$$h_i(t) = \frac{f_i(t)}{1 - F_i(t)} = \frac{\beta}{\alpha_i} \left(\frac{t}{\alpha_i} \right)^{\beta - 1} \tag{4.10}$$

where f is the probablity density function, F is the cumulative distribution function of the failure distribution and the index i denotes the respective stress level. However in this case, it is more useful to look at the reliability function R rather than the failure function F, which are complementary functions (see Eq. (2.1)).

$$R_i(t) = 1 - F_i(t) = \exp\left\{ -\left(\frac{t}{\alpha_i} \right)^{\beta} \right\} \tag{4.11}$$

With Eqq. (4.10) and (4.11), the relation of reliability function R and hazard function h can be established

$$R_i(t) = \exp\left(-\int_0^t h_i(x)\,\mathrm{d}x\right) \tag{4.12}$$

and combined with Eq. (4.9) to obtain the reliability function of the second stress step R_2 after the stress change time t_2:

$$\begin{aligned} R_2(t > t_2) &= \exp\left[-\left(\int_0^{t_2} h_1(x)\,\mathrm{d}x + \int_{t_2}^t k_2 \cdot h_1(x)\,\mathrm{d}x\right)\right] \\ &= R_1(t_2) \cdot \exp\left[-\left(\int_{t_2}^t k_2 \cdot h_1(x)\,\mathrm{d}x\right)\right] \\ &= R_1(t_2) \cdot \exp\left[-\left(\int_{t_2}^t h_1(x)\,\mathrm{d}x\right)\right]^{k_2} \\ &= R_1(t_2) \cdot \exp\left[-\left(\int_0^t h_1(x)\,\mathrm{d}x - \int_0^{t_2} h_1(x)\,\mathrm{d}x\right)\right]^{k_2} \\ &= R_1(t_2)\left[\frac{R_1(t)}{R_1(t_2)}\right]^{k_2} \end{aligned} \tag{4.13}$$

As a result, the first two stress steps of the TFR model are usually formulated like following:

$$R_{\mathrm{TFR}}(t) = \begin{cases} R_1(t), & t \leq t_2 \\ R_1(t_2)\left[\frac{R_1(t)}{R_1(t_2)}\right]^{k_2}, & t > t_2 \end{cases} \tag{4.14}$$

When extending the TFR model to more stress steps, like shown for R_3

$$\begin{aligned} R_3(t > t_3) &= R_2(t_3)\left[\frac{R_2(t)}{R_2(t_3)}\right]^{k_3} \\ &= R_1(t_2)\left[\frac{R_1(t_3)}{R_1(t_2)}\right]^{k_2}\left[\frac{R_1(t_2)\left[\frac{R_1(t)}{R_1(t_2)}\right]^{k_2}}{R_1(t_2)\left[\frac{R_1(t_3)}{R_1(t_2)}\right]^{k_2}}\right]^{k_3} \\ &= \frac{R_1(t_2)}{R_1(t_2)^{k_2}}\frac{R_1(t_3)^{k_2}}{R_1(t_3)^{k_2 k_3}}R_1(t)^{k_2 k_3} \end{aligned} \tag{4.15}$$

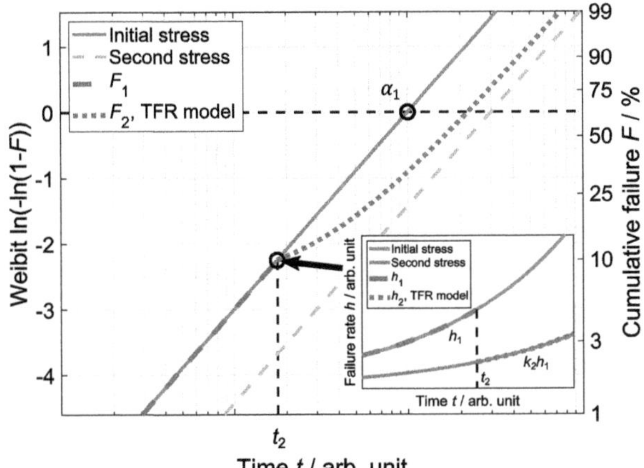

Figure 4.4: The TFR model utilizes a change in the initial failure rate as a means of modeling the CDF behavior of a successive stress step after the stress level change. As in the TRV model, the CDF of the second stress level is only considered in the model formulation as a means of determining the tampering coefficient k_i.

a pattern emerges which ultimately results in a multi-step-stress formulation of the TFR model

$$R_i(t) = \prod_{j=2}^{i} R_1(t_j)^{\prod_{l=1}^{j-1} k_l - \prod_{m=1}^{j} k_m} \cdot R_1(t)^{\prod_{j=1}^{i} k_j} \tag{4.16}$$

for $t_i < t \leq t_{i+1}$, where $k_1 = 1$ (compare [Mad93] for a similar expression). When deploying Weibull distributions for the application of the TFR model, the tampering coefficient k_i is defined by the constant ratio of the corresponding failure rates and can be correlated with Eqq. (4.4) and (4.8):

$$k_i = \frac{h_i}{h_{i-1}} = \left(\frac{\alpha_{i-1}}{\alpha_i}\right)^\beta = \left(\frac{1}{\mathrm{AF}_i}\right)^\beta = (\kappa_i)^\beta \tag{4.17}$$

As it turns out, similarly to the TRV model, the tampering coefficient k_2 is determined by the failure distribution F_2 of the second stress, but solely the first distribution F_1 is used for expressing the failure behavior after the stress change t_2. As illustrated in Fig. 4.4, the failure rate is changed from h_1 of the initial distribution to $k_2 h_1$ of the second distribution at t_2.

The course of the CDF is consequently not determined by any feature of the probability plot itself but rather by the failure rate plot in the inset of Fig. 4.4.

As demonstrated, the TFR model utilizes changes of the failure rate to describe the behavior of the CDF at stress level changes. This differs fundamentally from the previous cumulative damage models and hence results in a different failure prediction for varying step-stress levels.

4.2.4 Experimental Validation of Cumulative Damage Behavior

When elaborating the cumulative damage CE, TRV, and TFR model for typical Weibull failure distributions of thin SiO_2 dielectrics which exhibit Weibull slopes β with values around $\beta \approx 1$, the failure prediction after stress level change differs only marginally. Figure 4.5 illustrates these model differences in relation to the statistical variation of randomly generated artificial Weibull data. As a result of this empirical indistinguishability, initial testing was conducted for general validation of cumulative damage and with model predictions from the CE model, due to reasons of simplicity.

For this purpose, step-stress accelerated life testing (SSALT) measurements were performed on the simple MOS capacitor test structures that are depicted in Fig. 3.3a with an ultrathin (2.7 nm) insulating SiO_2 dielectric, and an Al contact pad. For the measurements, the semiconductor analyzer Keysight B1500A was used in constant voltage mode and applied positive voltages to the contact pad, resulting in the accumulation of the MOS-capacity as previously described in section 3.2.

The applied variable stressors in these experiments are electrical voltage and temperature of the device. Prior to the stress measurements, each device was tested with an initial integrity test to filter the specimens for yield and avoid left censored data. Thus, the leakage current of each capacitor was measured at 1.0 V to ensure oxide integrity. The failure criterion for the MOS capacitors was specified as a hard breakdown, defined as a local current jump of at least 1 µA at a sampling rate of 10 Hz. The individual recorded current–time characteristics are evaluated by means of analysis methods presented in chapter 2.2 and then cumulated and plotted in the Weibull probability plots of Fig. 4.6.

The voltage step-stress measurement in Fig. 4.6a where performed at room temperature. The lower and higher stress voltage were chosen as 2.5 V and 2.7 V, respectively. After stressing with either stress voltage, the stress level was changed after the median time t_{50} of the Weibull distribution and kept constant until a failure of the device occurred. The cumulative failure fraction of 50 % was chosen arbitrarily and the change point calculated beforehand. These times were $t_{50} = 569\,\mathrm{s}$ for the low and $t_{50} = 45\,\mathrm{s}$ for the high stress, which were obtained from the constant reference stress tests for both voltages. With respect to statistical variation, the failure distribution of the MOS capacitors follows the exact behavior of the prior calculated three-parametric Weibull distribution of Eqq. (4.2) and (4.5).

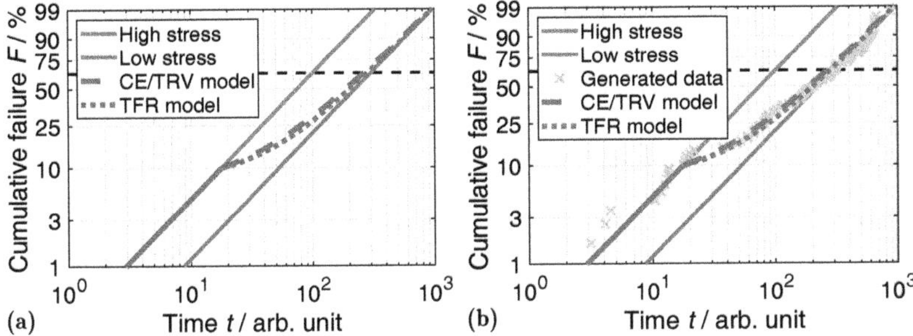

Figure 4.5: The difference of the CE, TRV, and TFR model is not distinctive when looking at model behaviors and artificial failure data. In both subfigures, Weibull distributions of constant stresses with slope $\beta = 1.3$ are marking high and low constant stress. The failure distributions of a 2-step-stress from high to low stress at the 10 % failure criteria are shown with blue lines. CE and TRV models coincide and are illustrated as a dashed line, the TFR model is shown as a dotted line. In subfigure **(b)**, orange crosses indicate 100 randomly generated Weibull data points of the depicted step-stress. Within the statistical variation of the artificial data, the cumulative damage models cannot precisely be distinguished.

The location parameter τ is in this case the difference of between both t_{50} times, which is $\tau = \pm 524\,\mathrm{s}$.

The temperature step-stress tests in Fig. 4.6b are conducted at constant voltage of 2.4 V and the two different temperature levels 50 °C and 80 °C. The median times for these stress levels were $t_{50} = 587\,\mathrm{s}$ for the low and $t_{50} = 90\,\mathrm{s}$ for the high temperature. Consequently, the respective location parameters are derived as $\tau = \pm 497\,\mathrm{s}$. The results show a comparable behavior of the cumulative damage measurements for temperature and voltage step-stresses and proof the general applicability of the presented cumulative damage models and the CE model in particular.

The described behavior does not only hold for two but also for more stress levels. Hence, SSALT measurements with three voltage stress levels have been performed for all possible permutation of these stress levels. Figure 4.7 depicts these reliability measurements with the voltage levels 2.5 V, 2.6 V, and 2.7 V and respective change times t_{20} and t_{63}. Divergent behavior to the prior calculated CE model prediction is mainly due to altered t_{20} and t_{63} times compared to the initial constant reference stress measurements and within the tolerable variation because of chip locations on the wafer due to oxide thickness variation and statistical effects.

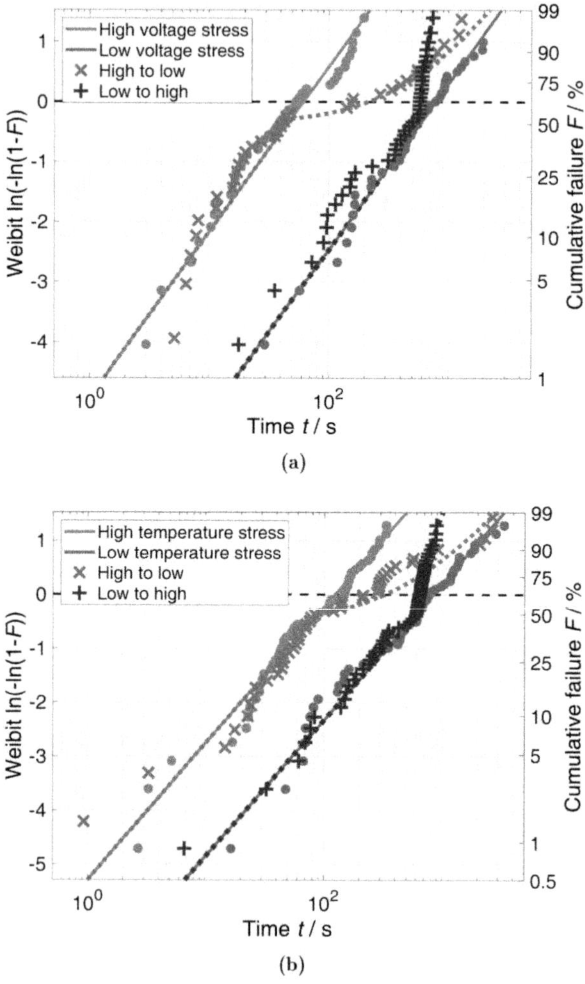

Figure 4.6: Breakdown distributions of MOS capacitors for two stress level permutations with the stress changed at the median time t_{50}. Crosses indicate the step sequence from high to low stress levels and pluses the vice versa sequence for **(a)** voltage and **(b)** temperature stress.

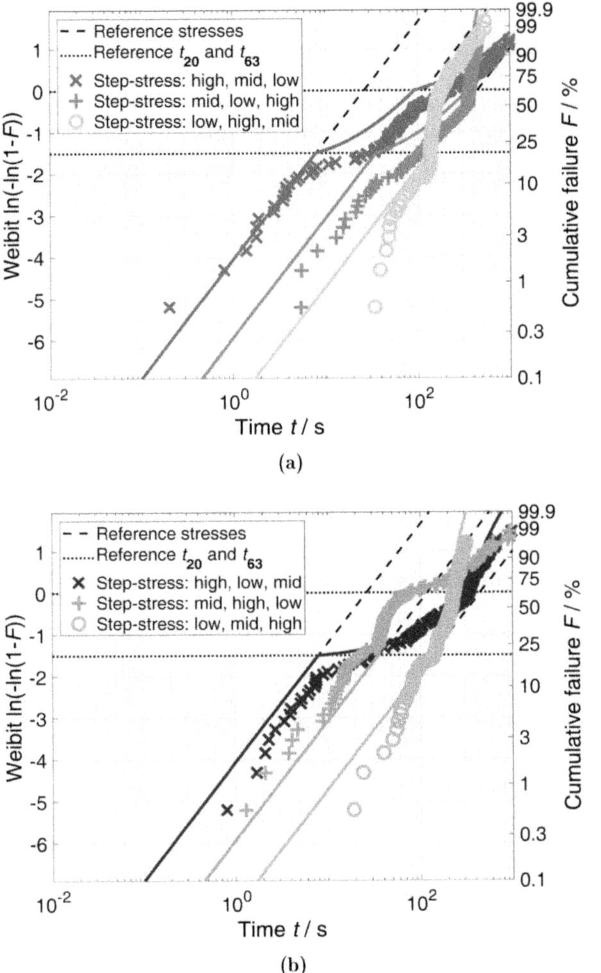

Figure 4.7: Step-stress data for all permutations of three different voltage stress levels. Stress change times are the respective t_{20} and t_{63} times of the reference stress measurements. Despite slightly different failure behavior of the reference CDFs and the first stress steps due to distant test-chip locations on the wafer, the SSALT data relate distinctly to the respective cumulative damage model prediction.

4.2.5 Model Comparison

After this first confirmation of cumulative damage behavior, a thorough comparison of the three aforementioned cumulative damage models reveals further mathematical and graphical similarities and differences. When elaborating Eqq. (4.1), (4.6), and (4.9) with Weibull distributions, the function of the first stress step is the unaltered Weibull CDF

$$F_1(t) = 1 - \exp\left\{-\left(\frac{t}{\alpha_1}\right)^\beta\right\}, \qquad t_1 < t \leq t_2 \tag{4.18}$$

whereas the formula of the second stress step $F_2(t)$ can be expressed accordingly to the chosen cumulative damage model

$$F_{2,\mathrm{CE}}(t) = 1 - \exp\left\{-\left[\frac{t - \left(t_2 - t_2'\right)}{\alpha_2}\right]^\beta\right\} \tag{4.19a}$$

$$F_{2,\mathrm{TRV}}(t) = 1 - \exp\left\{-\left[\frac{t_2 + \kappa_2\left(t - t_2\right)}{\alpha_1}\right]^\beta\right\} \tag{4.19b}$$

$$F_{2,\mathrm{TFR}}(t) = 1 - \exp\left\{-\left[\frac{t_2^\beta + k_2\left(t^\beta - t_2^\beta\right)}{\alpha_1^\beta}\right]\right\} \tag{4.19c}$$

for $t_2 < t \leq t_3$ [Wan04]. It is worth mentioning that the TRV and TFR model always refer to the scale parameter α_1 of the initial Weibull distribution, which is independent of the current stress step. On the contrary, every F_i of the CE model is specified by its respective α_i.

One finds that all models in Eq. (4.19) coincide if the Weibull slope $\beta = 1$, hence the distribution is an exponential one [Wan04, Bha89]. For $\beta \neq 1$, the TRV and TFR model cannot coincide for $\kappa_2 = k_2$ [Nab93], neither can the CE and TFR model. Additionally, as long as the failure distributions before and after the stress change belong to the same scale parametric family, the CE and TRV model coincide [Wan04, Bha89]. The term *scale parametric family* implies that only the scale parameter of the failure distribution is allowed to change between two consecutive stress levels. The type of distribution as well as any form or location parameter is considered unchanged. The requirement of scale parametric families is met for the commonly used Weibull and lognormal distributions, as they belong to log-location-scale families, under the premise of constant shape parameters [Sha14]. In this case, Eqq. (4.19a) and (4.19b) are reparameterizations of each other.

If in fact the Weibull slope β differs for two stress levels, the CE model would have the capability to model such a varying behavior by connecting two succeeding stress step distributions with the connection condition described in Eq. (4.1), whereas the

TRV model is restricted to the shape parameter of the initial stress distribution and cannot adapt to changing Weibull slopes. This difference between the CE and TRV model and its implications have not been discussed in the reliability community so far and would be an interesting topic for future research.

As already mentioned, it is difficult to distinguish between CE, TRV, and TFR model, especially when the Weibull slope β only differs little from 1 (see Fig. 4.5). In order to emphasize divergent failure behavior independently of the failure mechanism and underlying acceleration model we suggest to use alternating SSALT for two stress levels or so-called cyclical stress testing (CST) if additional stress levels are introduced.

As all three cumulative damage models coincide for $\beta = 1$, the resulting alternating SSALT failure behavior in Fig. 4.8 with a low and a high stress level is the same for all three models. As elaborated in the following sections, this converging behavior can be used to determine effective failure times for mission profiles.

Whereas the characteristics of the SSALT failure distributions in Fig. 4.8, with point-wise crossing and converging behavior of the alternating SSALT distributions for ever completed stress cycle, persists for the CE and TRV model also for $\beta \neq 1$ (see Figs. 4.9a and 4.9b), the TFR model deviates here. This is due to the commutative nature of the CE and TRV model in regards to the temporal sequence of stress steps and is a direct consequence of the definition of the acceleration factor in Eq. (4.4) and geometrical inferences of the Weibull probability plot. That is because for identical β, the acceleration factor of two Weibull distributions on a logarithmic time axis is independent of the cumulative failure F. Hence, stress changes form higher to lower and lower to higher stress levels are symmetrical. The TFR model does not have these properties as it alters the Weibull CDF in Eq. (4.18) for subsequent cumulative stresses into the non Weibull-like distribution of Eq. (4.19c) for $\beta \neq 1$.

For $\beta > 1$, the TFR model predicts two non-converging failure distributions which converge in a different way as the CE model (see Fig. 4.9a). Both distributions have a point of intercept during each stress cycle but end at different cumulated fails after every full stress cycle. The limits of this sequence were calculated numerically and feature shorter lifetimes than the predictions of the CE and TRV model. Another behavior can be observed for $\beta < 1$. As displayed in Fig. 4.9b, although both TFR SSALT failure distributions converge over time, they do not have any point of intercept. Contrary to the former case, the TFR model predicts longer lifetimes than the CE and TRV model for $\beta < 1$. This is due to the different emphasis of lower and higher stress levels by the tampering coefficient k_i in Eq. (4.17) compared with the acceleration factor AF_i.

As a consequence, the chronological order of stress levels is of high importance for reliability predictions using the TFR model. This is especially true when stress levels are not iterated like it is the case for alternating SSALT, since a converging behavior of different stress permutations is not likely for non-iterative stress sequences.

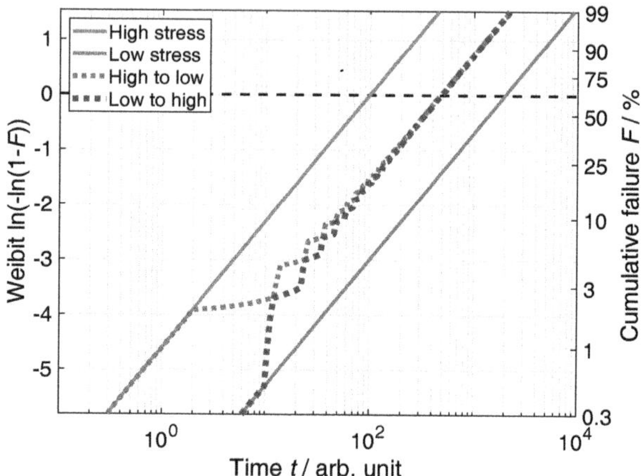

Figure 4.8: Illustration of an alternating SSALT of Weibull distributions with $\beta = 1.0$. Line and color coding is analogous to Fig. 4.6. Since all presented cumulative damage models coincide, only one dotted line is shown for each stress sequence.

The statements and conclusions drawn from these exemplaric illustrations for alternating SSALT with two stress levels remain also valid for cyclical step-stresses with multiple different stress levels as computer calculations show. With additional numbers and permutations of stress levels, the essential characteristics of the cumulative damage models persist, but plotted figures become increasingly less comprehensible.

For the successful usage of cumulative damage models in general, an acceleration model for the failure mechanism is useful but not required. The respective acceleration factors AF and tampering coefficients κ and k can be derived solely from empirical failure distributions under the considered stress levels. As for mission profile stresses, which usually exhibit various different stress levels, acquiring experimental data of every individual stress level — especially at normal operating conditions — is not feasible. Therefore, the Weibull parameters of the reference stress distributions are best obtained by verified acceleration models, like for example the exponential E model, power-law U model, and others for the failure mechanism of TDDB as presented in section 3.3 [JED16].

When focusing on the only result-based difference between the CE and TRV model, the computed behavior of alternating SSALT predictions of two stress levels with different Weibull slopes, as depicted in Fig. 4.15, shows only little deviations. These differences occur on the rather flat sections of the piecewise distribution function and thus are especially susceptible to experimental and statistical deviations of the plotted failure data.

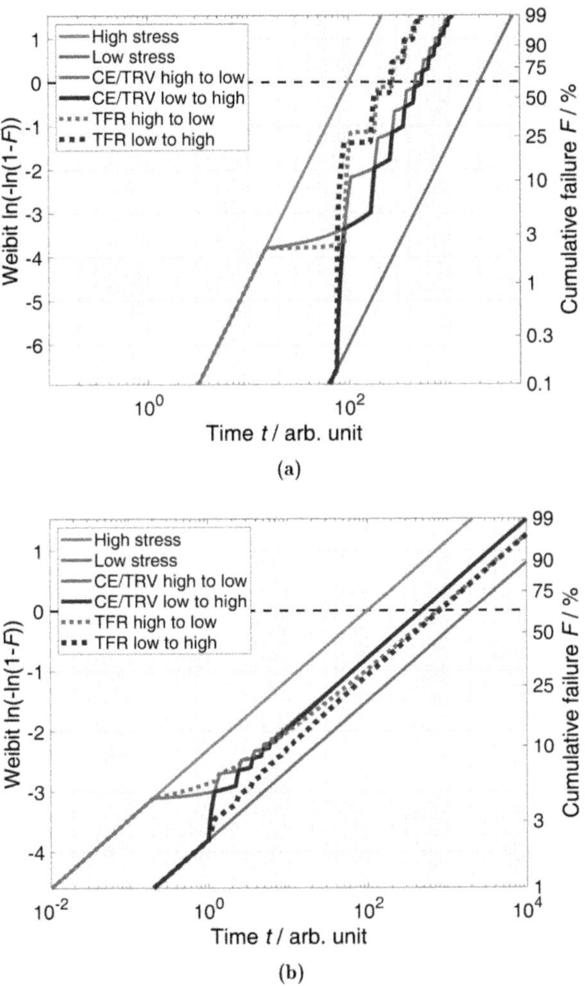

Figure 4.9: **(a)** For an alternating SSALT of Weibull distributions with $\beta = 2.0$, the TFR model predicts differing failure behavior. The first step-stress determines the subsequent failure trend, because the curves do not intersect at the end of each stress cycle, whereas the CE and TRV models exhibit failure characteristics analog to those in Fig. 4.8. **(b)** Weibull distributions with $\beta = 0.5$ for alternating SSALT lead to a converging but not intersecting failure behavior of the predictions for the TFR model with longer lifetimes than those for the CE and TRV model.

Due to this and the reason that different shape parameters are interpreted as indicator of altered failure mechanisms, this aspect of the CE model is systematically avoided and has not been previously investigated in literature.

Furthermore, as experiments on this kind of setup, i.e. strictly alternating stress levels or even decreasing stress levels for TDDB reliability tests, have not been reported in literature as well, the empirically observed cumulative damage behavior of the TDDB failure mechanism has also not been compared with different cumulative damage models so far. Hence, these open issues will be investigated among other things in the following sections on cumulative damage experiments.

4.3 Mission Profile Stress and Effective Stressors

The most important part of reliability assessment against mission profile requirements is the transformation of such mission profiles into reliability stress test conditions. The objective of this endeavor is to confirm an industry-wide used approach of deriving effective stresses from mission profile requirements, which has been the result of assumptions and best practice at that time but has not been empirically proven for semiconductor devices so far.

Since mission profiles are typically arranged in different diagram types, the necessary conditions of transforming or abstracting the information of mission profiles into stress test conditions have to be determined. Three main abstraction levels of mission profiles have been identified and illustrated in Fig. 4.10:

- The most detailed and unprocessed type of mission profile for a single individual stressor, like temperature or voltage, is the *stress time-series*, in which the time-dependent course of the stressor can be given in a simple stress–time diagram. As this kind of mission profile is quite detailed and challenging to obtain from real-life applications especially for the entire lifetime of products like cars due to measurement constrictions and also data privacy of the customer, they are rarely found among stated mission profile requirements.

- Far more common for industry applications are *stress histograms*, which are sometimes called *collective mission profiles*. These histograms show cumulated times for stress levels intervals, omitting any sequential component of the considered stressor. They are easier to use and describe than stress time-series and are better suited for resultant consolidations of a group of different applications or use cases in particular. Because of these characteristics, stress-histograms can be considered as the first abstraction level of a raw data mission profile.

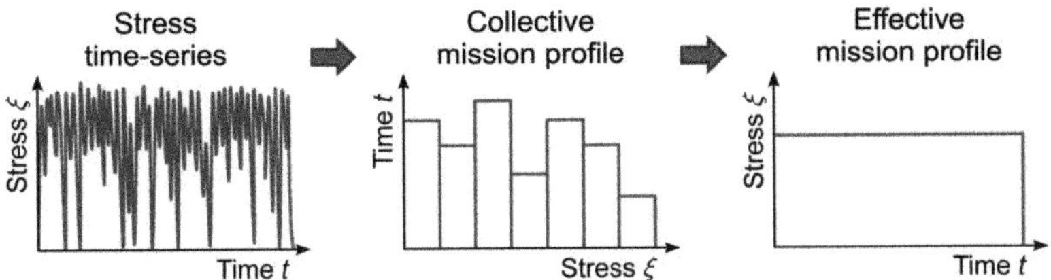

Figure 4.10: The conceptual abstraction levels of mission profiles.

- The conceptional most simplistic and further abstracted type is the *effective mission profile* which consists of a single effective stress level or effective stress time and a respective lifetime or stress level for a specific failure mechanism and device. This effective mission profile can be directly utilized as reference stress for reliability analysis and testing as described in section 2.2. Effective stresses are often preferred when comparing different mission profiles and usually computed in order to determine to what degree reliability requirements are covered by the tested or presumed reliability capability of a device.

The resulting leading question is: What conditions are necessary for the transformation of one abstraction level to the next one and are they satisfied for the successive experiments?

First, it has to be physically correct to divide a stressor into segments to be able to omit stressor dynamics such as transients.

Second, if in addition to the latter the sequence of segmented stress levels can be changed without any alteration of the failure physics, then it is possible to transform a stress time-series into a stress histogram. These criteria for the failure mechanism should also be satisfied in case a stress histogram is constructed artificially or merged from multiple mission profiles.

Finally, there has to be a physically justified method to obtain an effective stress from a given stress histogram. A mere arithmetic average might yield results that are not verifiable by experiments. If all the above requirements are fulfilled, then to an arbitrary failure rate the necessary effective stress level or testing time can be determined for a mission profile.

In our case, the experimental parameters are kept in an interval, in which the commutative nature of the cumulative damage model is assumed to be valid and the failure mechanism does not change for the range of applied stress levels. Therefore, the first two key questions can be answered affirmatively in the case of TDDB and a cumulative damage model that is commutative in regards to the stress sequence.

Resulting from section 4.2.5, the CE and TRV model are such models. Since both models are equivalent under these given circumstances, the mathematical formulation of the CE model is easier to apply, because of the existing recursive model formulas. Hence, when the CE model is mentioned in the course of the following sections, it also refers to the TRV model in analogous manner.

4.3.1 Transformation of Step-Stress into Effective Stress

In order to derive an effective stress level from a stress histogram, the representation of a stress histogram can be interpreted as a step-stress function. With a commutative stress model like the CE model, the different stress levels of such a histogram can be applied in any temporal sequence. The cumulated stress of different stress level permutations will be equivalent as long as the relative proportions of the different levels are unaltered. For the purpose of approximating realistic stress cycles and to expose test samples to counterbalancing stress, cyclical step-stress accelerated life testing (SSALT) is used as mission profile stress. The more full stress cycles with all stress levels are completed, the more points in time with equivalent stress budgets exist in the failure probability plot and the behavior of the effective failure distribution can be traced in more detail.

Without loss of generality, we can depict a cyclical SSALT with only two different stress levels in Fig. 4.11 and draw important inferences. When applying two arbitrary stress levels with fixed stress times in an alternating sequence, the CDFs of the two possible stress-permutations intersect periodically every full stress cycle, marking states of equal stress budgets. A closer look at these interceptions reveal that they are positioned in a constant arithmetic ratio to the initial failure distributions, which translates to a constant distance on the logarithmic time axis. As a direct consequence, all intersection points and therefore cumulative failure values of equivalent stress are placed on a straight line, parallel to the CDF of the initial reference stresses for geometrical reasons. This parallel failure distribution represents the distribution function of an equivalent constant effective stress of the used step-stress functions. This equivalent effective CDF signifies the distribution function of an unaltered failure mechanism with a scale-parameter — and therefore a stress level — that is in-between the two reference stresses. Additionally, all proportional step-stress periods will result into the same effective stress distribution.

In order to obtain a value for the effective stress level, the logarithmic distance, i.e. the ratio of the effective stress to a reference stress has to be determined. For the case of an unaltered failure mechanism, the acceleration factor AF (see Eq. (2.5)) between the stress levels is uniform along the cumulative failure axis. Therefore, the acceleration factors for multi-level SSALT can be defined by using the respective characteristic lifetimes α of different stress levels i and j:

$$\mathrm{AF}_i = \frac{\alpha_i}{\alpha_j} \tag{4.20}$$

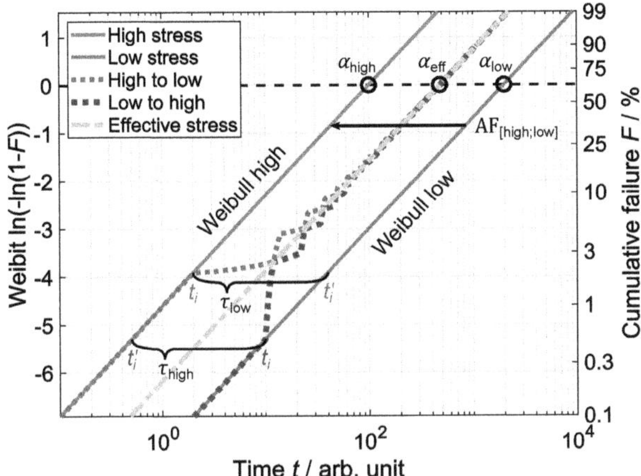

Figure 4.11: The different permutations of alternating stress levels, depicted as dotted lines, approach the CDF of a constant effective stress level, which is plotted as dot-dashed line, as more step-stress cycles are completed. This constant effective stress CDF can be determined with Eq. (4.22) by using the acceleration factor (AF) between the reference stress distributions and the respective stress times t_i.

With $\Delta t_i = t_{i+1} - t_i$ denoting the duration of the stress level step i, the CDF of the underlying failure distribution (see Eq. (2.6b)) can be utilized to express the ratio of the effective stress distribution to the sum of Δt_i over any number of complete stress cycles of the SSALT:

$$\frac{\alpha_{\text{eff}}}{\sum\limits_{i=1}^{n} \Delta t_i} = \frac{\alpha_j}{\sum\limits_{i=1}^{n} \frac{\Delta t_i}{\text{AF}_i}} \tag{4.21}$$

Switching from individual stress step times Δt_i to stress level proportions $p_i = \Delta t_i / \sum_{i=1}^{n} \Delta t_i$ enables a reformulation of Eq. (4.21) to give the effective acceleration factor of a mission profile in respect to an arbitrarily chosen reference stress level:

$$\text{AF}_{\text{eff}} = \frac{1}{\sum\limits_{i=1}^{n} \frac{p_i}{\text{AF}_i}} \tag{4.22}$$

Concluding from Eq. (4.22), the acceleration factor of an equivalent constant effective stress distribution of cyclical SSALT can be derived by the *harmonic mean* of the acceleration factors of the individual step-stress levels, weighted by the respective temporal proportions p_i of each stress level. This deduction of an effective stressor is independent of the acceleration law as it only utilizes the cumulative properties of the CE model and can also be used exclusively with empirically determined failure distributions.

Hence, a physically justified method has been represented to obtain an effective constant stress from a given stress histogram if the cumulative damage model can be considered commutative in regards to the stress level sequence. This also holds for mission profiles with more stress levels and multiple stressors as will be discussed and experimentally shown in section 4.4.

4.3.2 Experimental Validation of the Effective Stress Transformation

With the conclusion that a mission profile can be transformed into an equivalent effective stress level for the purpose of reliability testing and evaluation if certain criteria are met, the final experimental verification is still pending, even in literature. These criteria are namely that the stress can be divided into segments, the sequence of stress steps can be interchanged without altering the failure mechanism, and that an applicable effective stress can be determined. For the purpose of validating this stress transformation for the failure mechanism time-dependent dielectric breakdown (TDDB) for different semiconductor technology nodes and the individual stressors voltage and temperature, cyclical stress testing (CST) is performed. By applying this method of alternating stress levels, divergent failure behavior can be emphasized independently of failure mechanism and underlying acceleration model.

In order to validate the presented cumulative damage models for different devices, cumulative damage data are shown from MOS capacitors with ultrathin SiO_2 dielectric of 3.7 nm thickness (see Fig. 3.3a), which were fabricated on 4″-wafers in our university semiconductor cleanroom, as well as state-of-the-art transistors from GLOBALFOUNDRIES 22FDX® FD-SOI technology with high-k metal-gate stacks on 12″-wafers.

For the fully accelerated alternating SSALT experiment, firstly, two constant stress tests were conducted with different stress levels to act as reference measurements. The stress voltages at ambient temperatures and therefore the resulting stress levels are comparable to the overall failure acceleration that is typically used for wafer-level semiconductor reliability qualification by the industry. The resulting failure distributions exhibit nearly identical Weibull slopes and differ from each other by an acceleration factor of about 5, which is sufficient for this demonstration experiment.

Starting with those references, for each stress level in the alternating SSALT constant stress step times were determined. The ratio of these stress times was arbitrarily chosen in a way that the resulting CE model predictions were plotted approximately in the middle of the two reference measurements. Finally, CST is performed and the failure distributions are plotted onto the existing model simulations. Note that the presented cumulative damage models do not feature any independent fitting parameters. Their specific cumulative damage behavior is determined by no more than the Weibull parameters α_i and β of the reference measurements, the chosen stress times Δt_i, and depending on the specified model the acceleration factor AF_i or k_i, which is defined by the reference stress distributions. These given model parameters are unaltered when comparing empirical data with the different simulated model behaviors in Figs. 4.12, 4.13, and later in Fig. 4.14, because the beforehand calculated model predictions are likewise based on the reference measurements.

Although the TFR model does not feature commutative behavior regarding stress sequence for Weibull slopes $\beta \neq 1$ as discussed in section 4.2.5 and thus a transformation of a mission profile into a stress histogram might not be justified, the obtained failure data results will also be compared to TFR model predictions in order to investigate specifically the applicability of the TFR model for deriving effective stresses.

An aggravating factor of the tested samples is that the Weibull slope β of TDDB data is expected to usually be only slightly greater than 1, especially for ultrathin oxides [Som02] used in current leading-edge semiconductor technology nodes, and the predictions of the cumulative damage models might therefore not be distinguishable from each other. However, when examining the obtained failure data, both technologies show good intrinsic failure characteristics, which is supported by measured slopes β that are significantly greater than 1 [JED16]. This explicitly suggests matured technological processes and well-manufactured test samples.

For the ultrathin SiO_2 dielectric capacitors, the lower reference stress voltage was chosen to be 3.9 V and resulted in a characteristic lifetime $\alpha = t_{63}$ of 105.1 s, whereas the higher reference level with a voltage of 4.1 V had a characteristic lifetime of 18.9 s. Both Weibull distributions exhibit a slope of $\beta = 2.0$, which is comparatively high for such thin oxides, but allows for more delicate cumulative damage model distinctions. The stress step times were chosen to be 4.0 s for the lower stress and 2.0 s for the higher stress and were performed for 16 full stress cycles. The CE model simulation and calculation with Eq. (4.21) resulted in a predicted effective Weibull stress distribution with $\alpha_{\mathrm{eff}} = 41.7\,\mathrm{s}$ and $\beta = 2.0$. With the determined acceleration law and its derived model factors from section 3.3 for these test samples, the calculated effective stress level for this test setup is a voltage stress of 4.0 V. Analogous alternating voltage SSALT experiments have been conducted on GLOBALFOUNDRIES 22FDX® technology, but for intellectual property reasons, explicit failure times and corresponding stress levels are not disclosed.

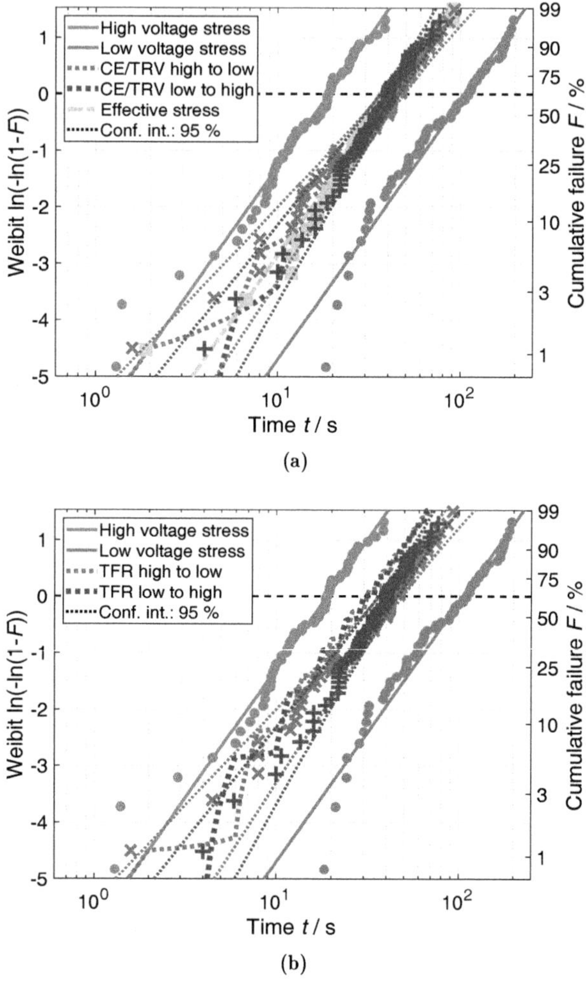

Figure 4.12: The experimental voltage CST data of MOS capacitors with a 3.7 nm thick SiO_2 dielectric confirm that **(a)** the experimental data coincide with the beforehand simulated failure behavior of the CE and TRV model. Additionally, the derived equivalent constant voltage stress overlaps with the corresponding alternating SSALT failure data. The experimental data also **(b)** falsify the failure prediction of the TFR model for the TDDB failure mechanism as the experimental data do not correlate to the simulated cumulative damage model behavior.

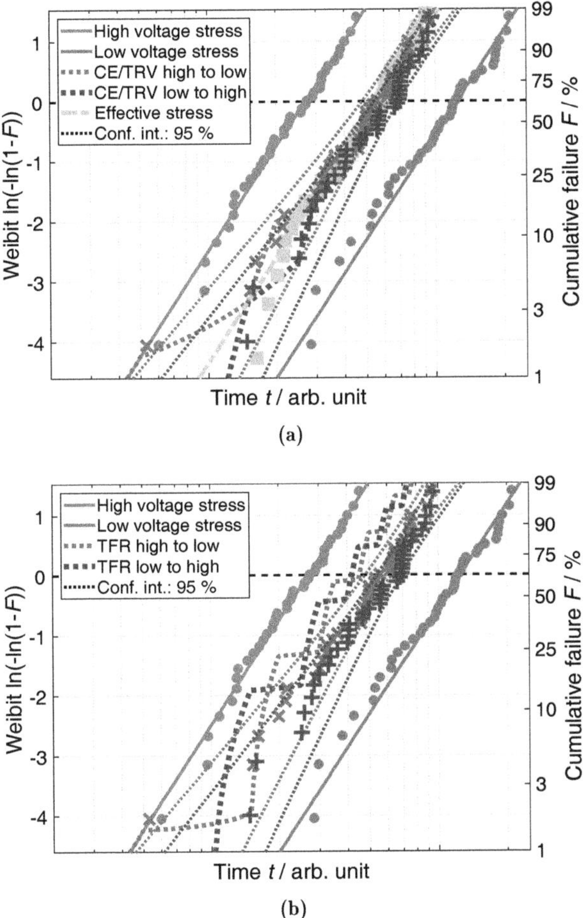

Figure 4.13: Merging of simulated and experimental voltage CST TDDB failure data of the 22FDX® technology from GLOBALFOUNDRIES. Subfigures **(a)** and **(b)** confirm the findings of Fig. 4.12 for this state-of-the-art semiconductor technology as well.

The results are depicted in Figs. 4.12a and 4.13a for the CE model and a clear coincidence of predicted model behavior and measured failure data is observed. This becomes apparent, when looking at the further progressed regions of the step-stress distributions. Especially the position of the characteristic lifetime t_{63} is a good point of comparison, because it is the pivot of the Weibull distribution and not susceptible to changes of the slope β (see Fig. 2.4a). In contrast, no difference between constant stress and the CST can be observed at low quantiles, because the stress levels are identical to the constant reference stress measurements or the failure behaviors have just recently diverged. Hence, with many repetitive stress cycles, the failure behavior of the CST measurements and predicted models will stabilize and expose existing model differences. Additionally, the first failure in an application is usually expected after a duration which is a few orders of magnitudes longer than the completion time of a typical mission profile cycle. By then, the different permutations of stress levels will have further converged and already coincide with an effective constant stress. The starting CDF of alternating stress levels would therefore not be observed in the field.

Due to the fact that the SSALT measurements are piecewise composed of two different Weibull distributions, they cannot be fitted easily, neither is the calculation of the confidence bounds any trivial. A deeper study on confidence bounds with Bayesian analysis and Markov chain Monte Carlo algorithms of the CE and TFR models is conducted in [Sha14]. The determination of exact confidence bounds was not a particularly topic of this work, so the here given confidence bounds are for the maximum likelihood estimation of SSALT measurements under the premise of a single Weibull distribution, which are more conservative than the exact confidence bounds and sufficient for this application.

On the other hand, no correlation of the TFR model prediction and the measured TDDB failure data of voltage stress can be found in Figs. 4.12b and 4.13b, respectively. The diverging characteristic of the individual TFR model stress permutations are not reproduced by the experimental data, nor is there a physical justification for the effective constant stress that corresponds to neither of the permutations.

Alternating temperature stress, as presented in Fig. 4.14, exhibits comparable cumulative damage behavior as voltage stress, but data from temperature SSALT are considerably more difficult to acquire by experiment than data from voltage SSALT. This is due to reasons of parasitic degradation of test samples because of mechanical wear and stress migration from temperature cycling that is independent from the here considered TDDB failure mechanism. In order to mitigate the influences of temperature dynamics, mature semiconductor technologies with remote and robust contact pads are needed, which are less susceptible to these types of interfering failure mechanisms and their degradation. Thus, marginally impaired temperature stress CST data could only be obtained for the GLOBALFOUNDRIES 22FDX® devices.

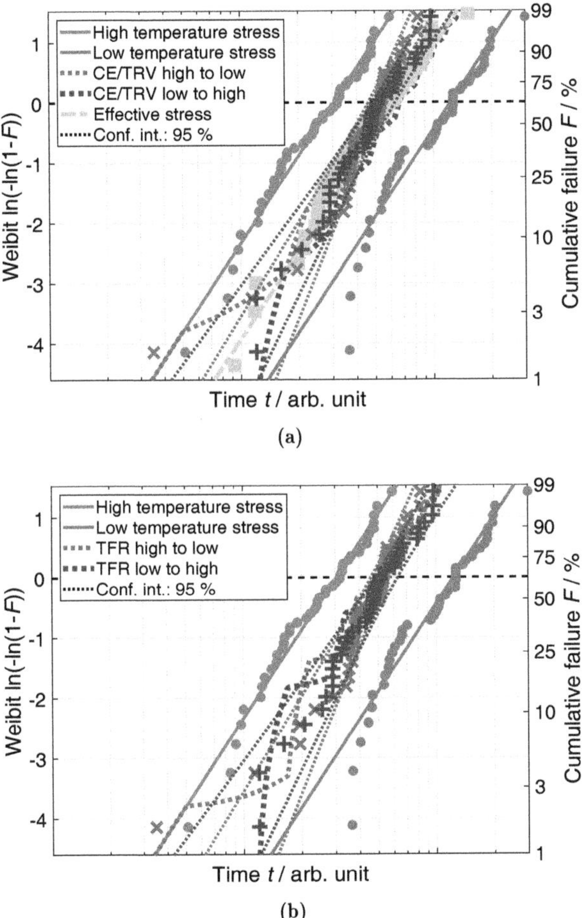

Figure 4.14: Experimental TDDB data of temperature CST on GLOBALFOUNDRIES 22FDX® transistors. **(a)** The data, which are affected by the parasitic degradation of temperature transients, agree with the beforehand simulated CE and TRV model behavior. Additionally, they approach and overlap the measured TDDB data of the calculated equivalent effective constant stress and confirm this step of abstraction for mission profile stresses. **(b)** The same experimental failure distributions also cross and overlap with the TFR model predictions, thus rendering the model distinction inconclusive for high-k semiconductor devices for this failure mechanism and test setup.

For that, the devices under test (DUTs) on the 12″-wafer were only heated locally by means of a custom made hot-air heating system during measurement. The temperature settings of the air heating system was calibrated on measurements of the temperature dependent reverse current of p–n diodes on wafer dices. During periods of heating up and cooling down, the specimens were electrically disconnected so that only a marginal amount of TDDB damage could have been accumulated during temperature changes, which are characterized by a saturating behavior. The changeover times between successive temperature levels were chosen long enough and the respective dwell time at each temperature level short enough to produce temperature steps which can be approximated to be constant.

As can be observed in Fig. 4.14, the CDFs of the temperature SSALT measurements show a particularly steeper slope than the reference measurements. This can be explained by the fact that the described parasitic degradation during temperature changes contributes to the reduction of lifetimes independently of the investigated TDDB which could not be completely avoided. However, this effect does not appear in the reference measurements, as well as in the effective constant temperature measurement, since these were not subjected to repeated temperature transients. For these reasons, although a cumulative damage behavior according to the CE and TRV model is shown to be analogous to voltage stresses, the TFR model cannot generally be ruled out, as the measured EDFs and the simulated model distributions cross over.

This could be remedied in future measurements by further suppressing parasitic degradation through the use of simplified and more robust test structures with shorter leads. Similarly, the constant stress measurements can also be designed as SSALT to be affected by lifetime degradation to the same extent.

The findings that the CE and TRV model for the failure mechanism TDDB and different consecutive temperature or voltage stress levels could be confirmed in this experimental setting are in line with the model of linear percolation of defects for the TDDB failure mechanism [Som02], which is commonly assumed for SiO_2 and high-k dielectrics and will result in a CE or TRV model behavior.

These outcomes confirm that it is valid for certain failure mechanisms like TDDB under voltage and temperature stress to transform mission profiles into an equivalent effective constant stress for reliability testing and evaluation. This requires solely a cumulative damage model, which is commutative in regards to the stress sequence, to be applicable and the failure distributions of the reference stresses by empirical or acceleration law analysis. Verifying the industry-wide used approach of using mission profiles for the reliability assessment of semiconductor components is of high importance, since established standard qualification processes for previously only constant stress requirements can also be used for the effective stress of mission profiles. The applicability of the CE and TRV model for TDDB will also be the central starting point for ramp-stress testing in the coming chapter 5.

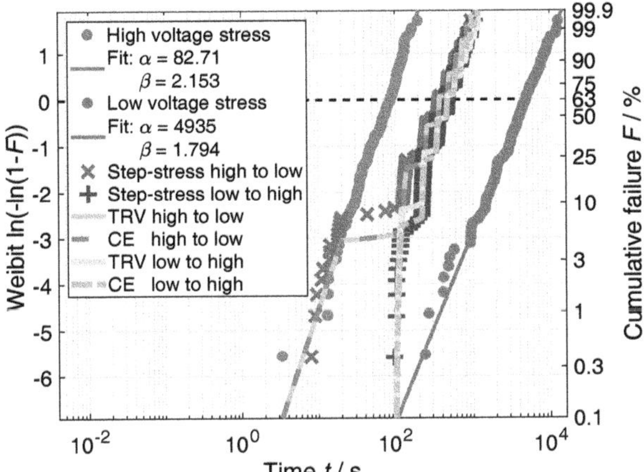

Figure 4.15: CE and TRV model predictions of CST with reference voltage stress distributions that exhibit different β slopes and experimental TDDB data. Whereas the CE model continues to result in converging step-stress CDFs, the TRV model exhibits diverging CST behavior that shows a parallel trend to the respective starting reference stress distribution. The slope difference $\Delta\beta$ of this experimental setup is found to not be large enough to justify further conclusions on the applicability of these cumulative damage models.

Apart from that, if a failure mechanism is governed in particular by its stressor dynamics, the Palmgren-Miner's rule with the rainflow-counting algorithm might be applicable to this section in an analogous manner. For that, one would only consider the stressor dynamics and substitute stress levels with transients as the main stress figure of merit, but this is not in the focus of this discourse.

Concluding from these findings and the importance of the emerging mission profile reliability requirements, cumulative damage model investigations on further failure mechanisms and semiconductor technologies are vital for reliable mission profile assessments. Additionally, it would be interesting for further studies to examine empirical differences between the CE and TRV model. As mentioned earlier in section 4.2.5, different sequences of stress regimes with varying Weibull slopes should yield further valuable insights into cumulative damage models.

Preliminary studies in Fig. 4.15 show failure data from a test wafer that features a relatively large difference in Weibull slopes $\Delta\beta = 0.36$ for stress measurements in different voltage stress regimes with a possibly altered failure mechanism.

For this case, the CE model is capable of taking reference stress distributions that feature different Weibull slopes in account and remain the characteristics of converging behavior over the course of many stress alternations and representing an effective stress CDF as described before. On the contrary, the TRV model has two separate step-stress distributions, which independently originate from the two reference distributions and are diverging in the same manner as both reference CDFs are diverging due to their Weibull slopes. Theoretical considerations show that the limit CDFs of the converging SSALT distributions are parallel Weibull CDFs to the corresponding reference stress behavior.

The depicted measurement data in Fig. 4.15 do match the presented characteristics of the CE slightly better than the calculated TRV model behavior, but as the model differences are only pronounced for the stress sequence starting with the lower stress and the SSALT data do not reproduce the cumulative failure percentage of the simulated stress step changes sufficiently well, the experimental outcome is not conclusive. To further examine this topic, new measurements with more widely varying Weibull slopes $\Delta\beta \approx 1$ would be desirable. In this context, also the interpretation of the limit CDF of CST as effective stress level needs to be reconsidered.

4.4 Multi-Dimensional Mission Profiles with Interdependent Stressors

As in section 4.3, in most cases of reliability qualification so far, mission profiles are considered to be one-dimensional, meaning only one stressor is examined while the remaining stressors are assumed to be constant. Whereas this may be applicable for basic circumstances, more complex requirements arise with the progress of automotive megatrends and the overall demand of higher performance and leading-edge technologies in new and expanding fields of application.

Multi-dimensional mission profiles, which consider two or more stressors and their interactions, as well as an increasing amount of operating modes with distinct environmental and functional loads are starting to become more common for reliability requirements. This is also a consequence of the above-mentioned megatrends in the automotive industry which extend the operating modes from operating and non-operating states to include operating modes like battery charging, vehicle preconditioning, standby for car connectivity and others.

The question arises, how to handle mission profiles in general and different operating modes in particular for semiconductor reliability. Literature on this topic is barely available and does not address the issues and needs of today's industrial reliability qualifications. Therefore, the theory of averaging acceleration factors and combining different stressors will be deduced and the reported concept of effective stress will be extended onto multi-dimensional mission profiles with combined stressors.

4.4.1 Multi-Dimensional Acceleration Factors

To start with, the acceleration factor (AF) is defined in Eq. (2.5) as the ratio of two different times-to-failure (TTFs) of a given component under normal operating (nop) and accelerated (acc) conditions. Hence, a resulting acceleration factor $\mathrm{AF}_{\mathrm{res}}$ can be determined for different stressors like temperature T, voltage U, and others. With that, the predicted lifetime can be calculated as:

$$\mathrm{TTF}_{\mathrm{nop}} = \mathrm{TTF}_{\mathrm{acc}} \cdot \mathrm{AF}_{\mathrm{res}} \left(T, U, \dots\right) \tag{4.23}$$

In general, this representation is only valid for degradations that do not change their failure mechanism or the shape parameter of their failure distribution for the entire range from normal operating to the highest stress conditions.

For general applicability, this $\mathrm{AF}_{\mathrm{res}}$ can be calculated with universal stressors X, Y, Z and their corresponding AFs. In case the individual stressors are given as single values, i.e. zero-dimensional mission profiles (see Fig. 4.16a), the resulting AF is calculated by the product of the individual contributions:

$$\mathrm{AF}_{\mathrm{res}} \left(X, Y, Z, \dots\right) = \mathrm{AF}_{X} \cdot \mathrm{AF}_{Y} \cdot \mathrm{AF}_{Z} \cdot \dots \tag{4.24}$$

If a one-dimensional mission profile that features different states i of one stressor, while additional stressors that are used to accelerate the life test are constant throughout the mission profile, like displayed in Fig. 4.16b, Eq. (4.24) can be modified to give the resulting acceleration factor as the multiplication of the constant AFs with the effective AF from Eq. (4.22):

$$\mathrm{AF}_{\mathrm{res}} = \frac{1}{\sum\limits_{i=1}^{n} \frac{p_{X_i}}{\mathrm{AF}_{X_i}}} \cdot \mathrm{AF}_{Y} \cdot \mathrm{AF}_{Z} \cdot \dots \tag{4.25}$$

The aforementioned case is the basic implementation that is used for the majority of mission profiles for qualification requirements and is being incorporated into standard qualification processes for the automotive industry. However, the more realistic scenario of mission profiles exhibits multi-dimensional mission profiles of different combined stressors (see Fig. 4.17a). Usually, these multiple stressors have interdependencies to other stressors, like temperature and relative humidity, or electrical current and temperature due to Joule heating. These dependencies cannot be ignored and must be taken into account when deriving the resulting acceleration factor of such a mission profile.

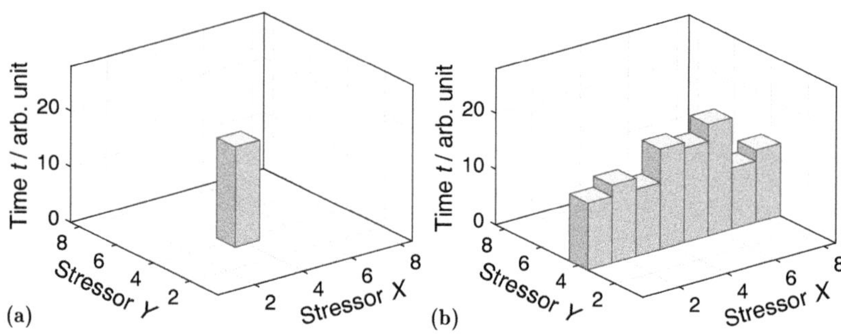

Figure 4.16: **(a)** Zero-dimensional mission profile: All stress values of use and accelerated conditions are constant. This used to be the long term standard for reliability requirements which consisted of static test conditions or a worst case estimation.
(b) One-dimensional mission profile: One stressor is represented as a stress histogram, whereas the other stressor values are static. This is the elementary form of a mission profile that is starting to be used more and more as qualification requirement for automotive components.

In order to test such multi-dimensional mission profiles like in Fig. 4.17a, an elementary four-state mission profile consisting of two stressors is chosen as the working example for reasons of simplification. The DUTs for this experimental setup are the elaborated MOS capacitors with remote contact pad and gaurd ring, as depicted in Fig. 3.3b. These were fabricated on 4″-wafers in the university cleanrooms with an marginal thicker ultrathin (4 nm) SiO_2 dielectric. The reference stresses 1 and 2 feature a high temperature, whereas 3 and 4 are subject to ambient temperature. Varied to that, the reference stresses 1 and 3 are stressed under higher voltage than the measurements 2 and 4. The respective stress step times of the CST are $\Delta t_1 = 4\,\mathrm{s}$, $\Delta t_2 = 10\,\mathrm{s}$, $\Delta t_3 = 2\,\mathrm{s}$, and $\Delta t_4 = 30\,\mathrm{s}$. These different stress states are then iterated until failure of the specimens, which are predicted to exhibit failure behavior according to the CE model.

Because the concept of effective stress in accordance with Eq. (4.22) is quite compelling for all forms of mission profiles, care has to be taken that stressor interdependencies are not neglected. Therefore, when trying to derive the resulting acceleration factor just solely by multiplication of independent $\mathrm{AF}_{\mathrm{eff}}$ as follows

$$\mathrm{AF}_{\mathrm{res}} = \frac{1}{\sum\limits_{i=1}^{n} \frac{p_{X_i}}{\mathrm{AF}_{X_i}}} \cdot \frac{1}{\sum\limits_{j=1}^{m} \frac{p_{Y_j}}{\mathrm{AF}_{Y_j}}} \cdot \frac{1}{\sum\limits_{k=1}^{o} \frac{p_{Z_k}}{\mathrm{AF}_{Z_k}}} \cdot \ldots \tag{4.26}$$

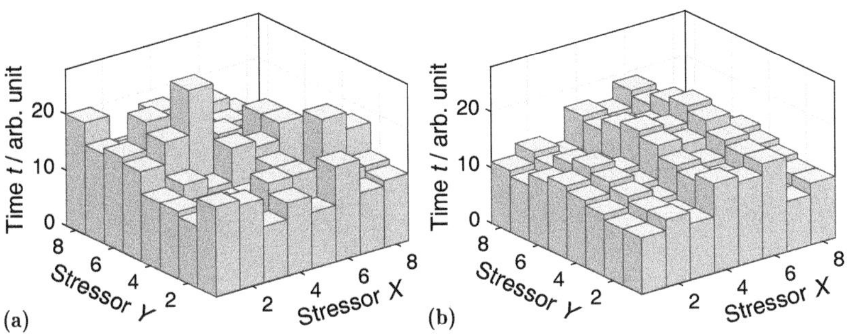

Figure 4.17: **(a)** Multi-dimensional mission profile: All stressors can exhibit stress distributions that are often illustrated as histograms and are possibly independent or coupled and feature interdependencies with other stressors. This is a more detailed and realistic approach to reliability requirements which, on the contrary, requires a deeper understanding of underlying mechanisms and increased effort for the composition of such mission profiles.
(b) Special case of a multi-dimensional mission profile: Although various stressors feature stress histograms, their stress distributions are not interdependent. Therefore, a resulting acceleration factor can be derived in a more basic manner than with the mission profile in subfigure (a).

the red 'erroneous' effective stress in Fig. 4.18 would be obtained. This failure distribution does not comply with the experimental iterating step-stress data, especially not at t_{63}, the pivot of the Weibull distribution. Moreover, the t_{63} of the 'erroneous' effective stress distribution is not enclosed within the 95 % confidence interval of the experimental step-stress data.

In order to consider coupled and interdependent stressors, the harmonic mean (see Eq. (4.22)) must include all relevant stressors:

$$\mathrm{AF}_{\mathrm{res}} = \frac{1}{\sum\limits_{i=1}^{n}\sum\limits_{j=1}^{m}\sum\limits_{k=1}^{o} \frac{p_{i,j,k}}{\mathrm{AF}_{X_i}\cdot\mathrm{AF}_{Y_j}\cdot\mathrm{AF}_{Z_k}}} \tag{4.27}$$

This creates $n \cdot m \cdot o$ composed stress level states that have time percentages $p_{i,j,k}$ which are derived by:

$$p_{i,j,k} = p_{X_i} \cdot p_{Y_j} \cdot p_{Z_k} \tag{4.28}$$

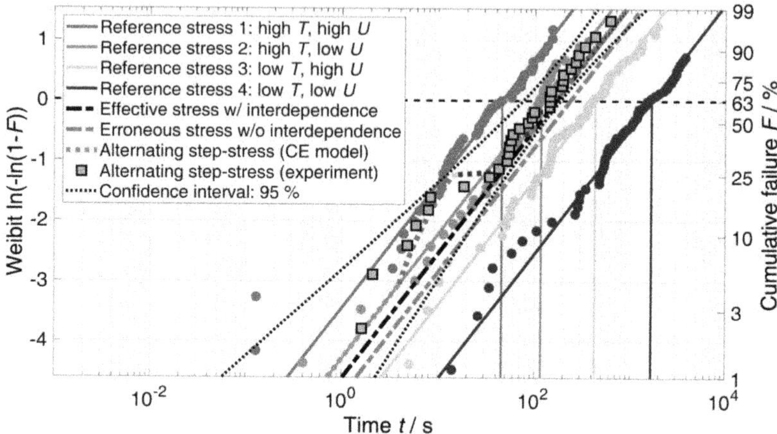

Figure 4.18: Experimental TDDB data of iterating step-stress measurement, consisting of coupled temperature and voltage stresses, confirm the in beforehand calculated failure behavior of the CE model. The 95 % confidence interval of the step-stress data confirms the equivalence of the step-stress data and the derived effective stress according to Eq. (4.27) and falsifies the resulting acceleration factor from Eq. (4.26) for this application, as the interdependencies of the stressors are neglected in the latter equation.

When looking at the experimental failure data in Fig. 4.18, it becomes apparent that the iterating step-stress data comply with the CE model prediction and converges with the equivalent effective stress distribution determined by $\mathrm{AF}_{\mathrm{res}}$ from Eq. (4.27) within statistical variations.

Concluding for coupled stressors temperature T and voltage U, the resulting acceleration factor of Eq. (4.27) does not equal Eq. (4.26):

$$\mathrm{AF}_{\mathrm{res}} = \frac{1}{\sum\limits_{i=1}^{n} \frac{p_i}{\mathrm{AF}_{T_i} \cdot \mathrm{AF}_{U_i}}} \neq \frac{1}{\sum\limits_{i=1}^{n} \frac{p_i}{\mathrm{AF}_{T_i}}} \cdot \frac{1}{\sum\limits_{i=1}^{n} \frac{p_i}{\mathrm{AF}_{U_i}}} \tag{4.29}$$

But while Eq. (4.27) creates a high number of composed states, especially for highly detailed mission profiles, mission profiles with complete independent stressors can be processed with significantly less effort. In these cases, stressors with independent stress–time distributions, as depicted in Fig. 4.17b, can be treated as independent effective stresses and allow an evaluation with Eq. (4.26).

4.4.2 Inferences from Interdependent Stressors

As demonstrated on a basic four-state step-stress mission profile of interdependent stressors, the impact of neglecting those dependencies can lead to different resulting acceleration factors and therefore to diverging time-to-failure (TTF) estimations.

In order to illustrate the impact dimension that negligence of stressor dependencies can have for reliability qualification, three different real-world mission profiles for the same automotive application from AUDI are presented in Fig. 4.19. These are temperature–relative humidity mission profiles for organic light-emitting diode (OLED) vehicle lighting, covering passive and active operating stresses with self-heating effects for a presumed required lifetime of 15 years. For the observed OLED failure mechanism, the Peck model [Pec86] is used to calculate the individual acceleration factors

$$\mathrm{AF} = \left(\frac{\mathrm{RH}_{\mathrm{nop}}}{\mathrm{RH}_{\mathrm{acc}}}\right)^{n} \cdot \exp\left[\frac{Q_{\mathrm{a}}}{k_{\mathrm{B}}}\left(\frac{1}{T_{\mathrm{nop}}} - \frac{1}{T_{\mathrm{acc}}}\right)\right] \tag{4.30}$$

from normal operating $\mathrm{RH}_{\mathrm{nop}}$, T_{nop} and accelerated conditions $\mathrm{RH}_{\mathrm{acc}}$, T_{acc}, where RH is the value of relative humidity, T is the temperature in Kelvin, n is the Peck exponent (which is usually negative), and Q_{a} and k_{B} denote the activation energy and the Boltzmann constant. A standard 85 °C/85 % RH accelerated lifetime test is used as reference stress condition.

Hence, one might start straightforward with calculating the effective temperature and effective relative humidity of the given mission profiles in respect to the effective acceleration factors by means of Eq. (4.22). This would reveal not much of a difference between all three mission profiles (cf. Table 4.1). With that, also the $\mathrm{AF}_{\mathrm{res}}$ by means of Eq. (4.26) would hardly differ. As a result, the accelerated lifetime test to ensure the required lifetime of 15 years would last between 12 135 h for Kuala Lumpur and 13 229 h for Tampa. That equals up to one and a half years of stress testing. Whereas the correct effective test times with Eq. (4.27) are 1456 h for Kuala Lumpur, 967 h for Tampa, and only 432 h for Dubrovnik (according to $\mathrm{AF}_{\mathrm{res}}$ from Table 4.1).

This does not only lead to considerably less time and money spent for testing and decreased time of development and validation cycles but is also critical for the qualification testing itself. Because in the case of the Dubrovnik mission profile, a 30-times increased test time is in most cases unlikely to yield sufficient reliability results if the technology is not fully matured. Hence, deriving the correct corresponding test times is especially critical for the decision of using new leading-edge technologies instead of older and mature ones for upcoming applications with a particularly high performance need, like autonomous driving or automotive electrification.

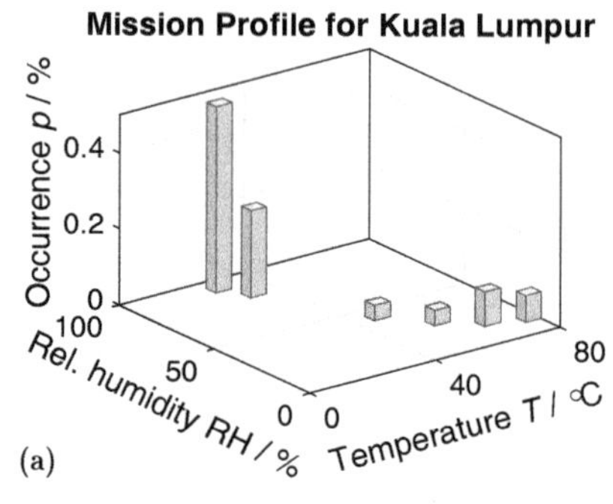

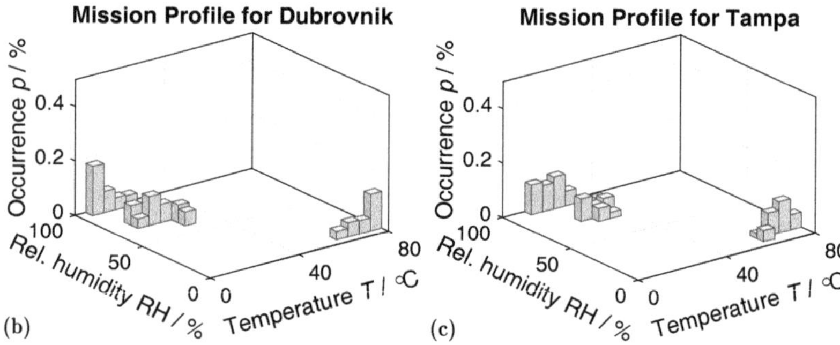

Figure 4.19: Mission profiles from AUDI for OLED vehicle lighting applications for different geographical locations in a two-dimensional stress histogram: **(a)** Kuala Lumpur, Malaysia, **(b)** Dubrovnik, Croatia, and **(c)** Tampa, Florida, USA. The relevant stressors for the regarded failure mechanism are temperature and relative humidity. The depicted occurrences are cumulative for active and passive loads.

Table 4.1: Acceleration factors for standard 85 °C/85 % RH lifetime test.

Mission Profile	$T_{\text{eff}}/\text{RH}_{\text{eff}}$	AF_{res} (Eq. (4.26))	AF_{res} (Eq. (4.27))
Kuala Lumpur	54.1 °C/77.9 %	10.8	**90.3**
Dubrovnik	55.1 °C/77.1 %	10.3	**304.3**
Tampa	54.7 °C/78.9 %	9.9	**135.9**

To illustrate the effects of interdependent stressors even further and give adaptable calculation paths for multi-dimensional mission profiles, an example calculation can be found in the appendix, section A.2.

In summary, mission profiles are becoming an important and necessary part of reliability requirements and qualification, as the automotive industry depends on new leading-edge semiconductor technologies in order to realize innovative applications without sacrificing the established reliability standards of their products. While basic implementations of mission profiles are handled more and more often in daily routine, more sophisticated requirements are entering the field of reliability qualification.

One aspect that is rarely brought to attention is the interdependency of relevant stressors in mission profiles. Hence, this work investigates the types of mission profiles and the methods of calculating effective and resulting acceleration factors for various types of independent and combined stressors in mission profiles. For this purpose, the first reported TDDB failure measurement of an iterating step-stress test of combined temperature and voltage stresses, to our knowledge, was conducted and presented. Thereby, it becomes apparent that taking stressor interdependencies into account has a significant impact on reliability predictions.

The dimension of this impact is illustrated by the comparison of three real-world mission profiles for automotive applications. It is concluded that neglecting stressor coupling in multi-dimensional mission profiles can have severe consequences on the reliability qualification process. This reaches from not satisfying the pursued reliability objectives for a given technology to not being able to utilize leading-edge technologies for novel innovative applications at all.

4.5 Technology Black Box

With the findings of this chapter, it was possible to develop the concept of a universal reliability calculator to support the supply chain of automotive OEMs in facilitating the processing of various different customer-specific mission profile requirements for novel and leading-edge semiconductor technologies within the scope of the "autoSWIFT" project. It was named Technology Black Box (TBB).

This reliability assessment tool is particularly characterized by the ability to be used throughout the technology and product development process to enable simultaneous engineering of the final product on different tiers in the supply chain, leading to highly reduced time to market, significantly mitigated risks of ultimately failing the reliability qualifications, and avoiding costly over-engineering. This is due to the fact that possible reliability issues can be identified in an early development phase and solved over the course of the development progress. For this, reliability evaluations are iterated with constantly refining information on technology parameters and reliability behavior. Likewise, the reliability capabilities are assessed against preliminary and constantly evolving mission profile requirements. This ensures that emerging problems and discrepancies are detected early on and can be addressed accordingly. Consequently, changes can be made to either the requirements by altering application times and loads, installation locations, cooling concepts, etc. or by adapting the product early in the development process when changes can be implemented more easily than later on.

To enable these functionalities, the prior presented method of transforming mission profile stress requirements into effective stress levels for comparison with available technology and product reliability data on basis of the CE and TRV model and cyclical SSALT is essential. Such is the extension of this method onto multi-dimensional mission profiles, as this improves the TBB's accuracy of reliability predictions even further.

As illustrated in Fig. 4.20, the TBB processes the input consisting of reliability requirements in the form of single stressors or multi-dimensional mission profiles, failure rate targets, application duty cycles and others with the required lifetime specified by the OEM or another tier. In order to do so, it is necessary for the technology supplier to enter the required technology parameter and reliability data and models of the respective devices for all relevant failure mechanisms into the so-called "Technology card" and the chip designer to state the planned chip design targets and dimensions. With this, the TBB compares the reliability requirements of the application with the reliability capabilities of the technology or product and outputs the degree of coverage in form of a red, yellow, or green 'traffic light' for the entire assessment. For a more detailed analysis of individual failure mechanism and device type different levels of detail can be displayed. This enables to discriminate critical failure mechanisms for planned applications and allows to find tailored solutions for the specific requirements of different mission profiles.

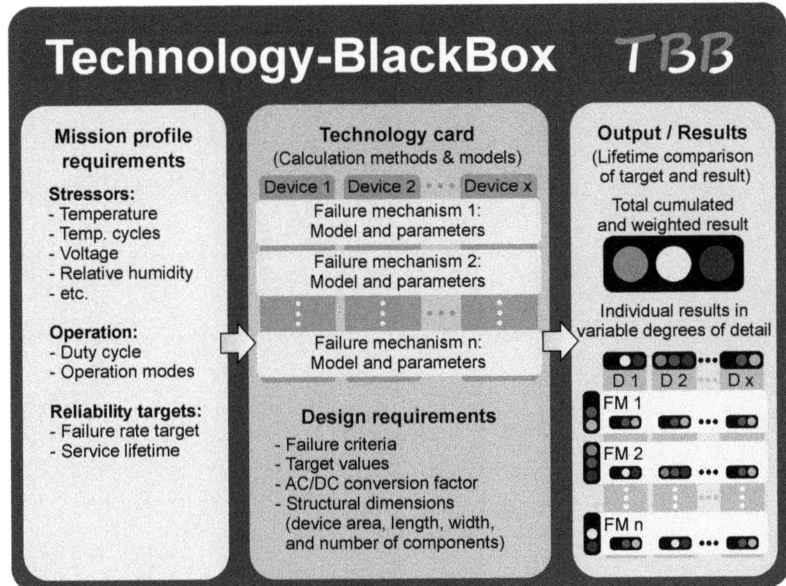

Figure 4.20: Schematic of the Technology Black Box, a prototype reliability evaluation tool which includes the findings of chapter 4, like assessment of multiple failure mechanisms against mission profile stress requirements with single or interdependent stressors, and other features for the first time.

Although the Technology Black Box prototype was initially focused on MOS technology and respective failure mechanisms, it is open for expansion to other semiconductor technologies as well. Also, further developed versions of the TBB have already been tested and used for the reliability evaluation of leading-edge technologies for novel automotive applications and have proven to be of great benefit for the manufacturers involved.

5 Elaborated Life-Testing with Ramp-Stress Tests

Semiconductor manufacturers, who produce semiconductor devices and components are faced with the problem of high reliability requirements put on them by their customers and the subsequent supply chain, an enormous output of semiconductor chips at wafer-level, which have to be tested at least to a certain degree, and are often very limited by time per test. The methods to master these challenging aspects are often summarized under the description of fast wafer level reliability (fWLR). The key essence of fWLR is the limited reliability testing time per chip. In order to solve this issue, more elaborate testing methods than the basic CVS have to be applied. Initially, step-stress accelerated life testing (SSALT) was utilized to decrease the required test time, ensure certain failure of the reliability test device for better data coverage, and avoid right censored data (see section 2.1.4).

The next logical steps to further improve SSALT is to reduce step time and step size, and thus increase the number of measurement steps for the purpose of avoiding grouped data. With further decreasing step sizes, this leads to an approximated continuous stress ramp. Due to the fact that voltage is one of the easiest to use stressors that can be controlled precisely and fast in a ramp shape manner and also accelerates quite many different failure mechanisms, ramp voltage stress (RVS) at ambient or elevated temperatures is an frequently used fWLR technique.

For the purpose of investigating ramp-stresses, the results on cumulative damage models of chapter 4 will be used and applied on voltage stress ramps and their transformation into constant voltage stress data for four different TDDB voltage acceleration models. A closer look at the transformation characteristics, the equivalence of stress between RVS and CVS, and the behavior of the RVS CDFs will help to analyze ramp-stresses and draw inferences from these findings for reliability evaluation with fWLR and further topics like acceleration model fitting.

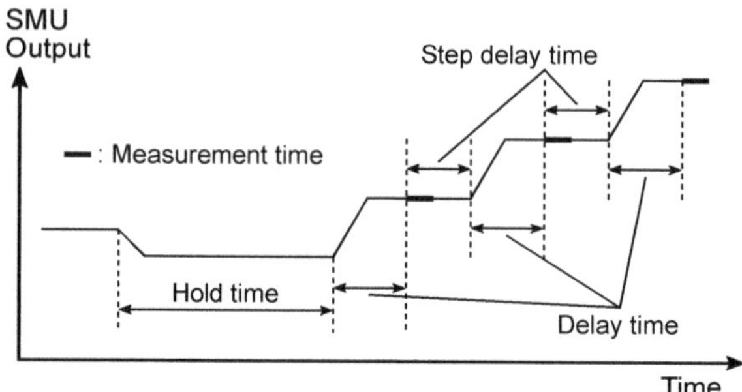

Figure 5.1: Schematic of the Keysight B1500A measurement time, delay time, step delay time, and hold time configuration.

5.1 Ramp Measurement Methodology

When conducting ramp-stress measurements, it is largely preferred to maintain a constant ramp rate in order to utilize the formulas derived in section 5.2 for low-effort application. With the knowledge of the composition of a stress step with measurement time and various delay times of the Keysight B1500A in Fig. 5.1 the measurement setup can be adjusted accordingly. Caution has to be taken as the measurement time varies for each measurement point and a possible range change introduces a very long delay before, during, or after the respective measurement.

In order to resolve these influences on the ramp rate, the step delay time can be utilized and adjusted to a value that is large enough to cover the occurring measurement time variations. Since the delay time setting is only available for remote control but not when controlling the B1500A via the Keysight program EasyExpert, constant measurement integration times have to be enforced when possible and accuracy settings have to be reviewed in order to not override prior integration time settings. Furthermore, the range change is added to the measurement overhead time, which cannot be specified directly, and can only be compensated by the step delay time for very low ramp rates. Hence, range changes are avoided by setting a fixed measurement range which is sufficient to cover the maximum expected current value of the ramp measurement, including the dielectric breakdown current, and is therefore often chosen to equate with the current compliance limit of the source/monitor unit (SMU). [Key16, Key17]

After achieving a constant ramp setting for the parametric measurement instrument, the ramp rate $\lambda = \Delta U/\Delta t$ can be adjusted by increasing delay and step delay times resulting in a larger Δt, and by changing the step size for a different ΔU.

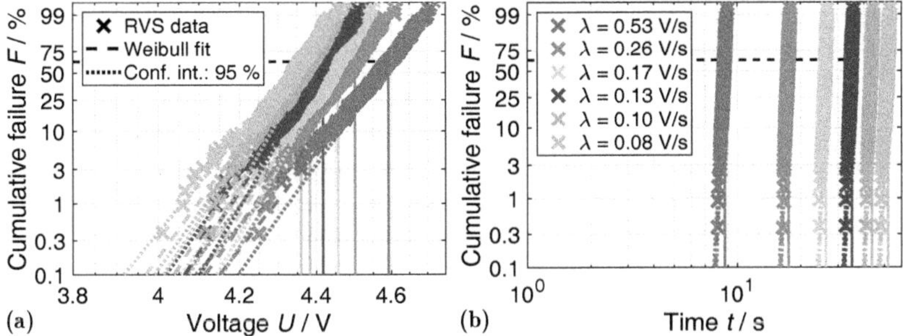

Figure 5.2: Experimental ramp-stress Weibull data with ramp rates λ from 0.08 V/s (light blue) to 0.53 V/s (dark blue) are usually plotted in a Weibull probability plot as **(a)** stress-to-failure plot with logarithmic voltage x-axis, **(b)** but can also be plotted as TTF plot with logarithmic time x-axis. Same colors indicate identical measurements and legend entries apply to both subfigures.

The recommended settings by the JEDEC for a pre-characterization test, which represents a fast ramp setup and functions as a point of reference for TDDB ramp tests, are given in the standards JESD92 [JED03] and JESD35 [JED01]:

“

- Ramp Rate: 1 MV/cm-s
- Step Height: not to exceed 0.1 MV/cm
- Voltage Step Time: less than 0.1 s
- Current Measurement Interval: minimum once per step.
- Maximum Voltage: 30 MV/cm

”

[JED03]

It is recommended to perform such ramp tests with the DUTs in accumulation bias in order to avoid inversion capacitance effects [JED01]. Analogously to CVS test setups, all E field stresses have to be corrected for at least the flatband voltage when measuring in accumulation and evaluating acceleration models that are dependent on the electrical field [JED03]. Initially, the ramp rate was specified as 0.1–1.0 MV/(cm·s) but has been changed to only one value to facilitate the reproducibility of voltage ramp tests performed accordingly to the JEDEC standards [Sue93]. This is due to the fact that breakdown voltage and breakdown time depend on the ramp rate as can be seen in Fig. 5.2.

The experimental ramp-stress measurements for this chapter are performed as SSALT with constant ramp rates on MOS capacitors with a 3.7 nm thick SiO_2 dielectric, like depicted in Fig. 3.3a, that were fabricated in the university cleanroom on 4″-wafers. For our measurements, the ramp rate λ is in the range from 0.08 V/s to 0.53 V/s, equaling 0.22 MV/(cm·s) to 1.43 MV/(cm·s), a constant step size of 5 mV, i.e. 0.014 MV/cm, and a maximum step time of < 0.06 s for the lowest ramp rate. With this, the experimental setup is within the boundaries of the recommended RVS parameters of the JEDEC, except for the highest ramp rate which alone exceeds 1.0 MV/(cm·s).

Due to the asymmetrically distributed breakdowns, data from RVS measurements are often assumed to be Weibull-like and are therefore treated as such. For this reason, RVS failure distributions are displayed in Weibull probability plots, like in Fig. 5.2, and can be plotted as stress-to-failure or analogously to CVS data as TTF diagrams. The sequence of measurements in Fig. 5.2a is inverted to that in Fig. 5.2b, because the specimens with the highest ramp rate will breakdown in the shortest time while achieving the highest breakdown voltages and vice versa. Furthermore, it becomes apparent that the stress-to-failure plot in Fig. 5.2a is more suitable for visual inspection of the RVS failure data than the time-to-failure plot in Fig. 5.2b. This is because the EDFs of measurements with different ramp-stresses are not as widely spread on the x-axis and therefore more characteristics and features of the distributions are noticeable. For that, stress-to-failure plots are the most common type of ramp-stress diagrams especially for standardized ramp-stress test parameters.

While RVS failure data are plotted mainly for inspection and comparative assessment of measurements, the more interesting object of study is the transformation of failures under ramp-stresses to constant stress conditions. This enables predictions about failure rates and the end of life of semiconductor components under typical use conditions. For this purpose, the effective stress times for RVS tests have to be deduced.

5.2 Effective Stress Times

As elaborated and shown in chapter 4, the CE model can be applied for TDDB and related failure mechanisms when dealing with cumulative damage. Since ramp-stress tests are a special type of stress sequence, in particular a monotonically increasing function of stress levels, usually a step-stress function or a linear function, the same mathematical structures as in chapter 4.2 can be utilized for this endeavor.

Generally for ramp-stress testing, a stress $\xi(t)$ is assumed that is variable over time. In order to relate a failure or breakdown of the DUT, indicated by the subscript BD, at stress level ξ_{BD} after a time t_{BD} to the reference test conditions or use conditions of the device, the acceleration factor (AF) (see Eq. (2.5)) can be utilized to describe a differential element of a reference stress time $\mathrm{d}t_{\mathrm{ref}}$ with a differential element of the observed time $\mathrm{d}t$. Subsequently, the equation can be rearranged and integrated:

$$\mathrm{d}t_{\mathrm{ref}} = \mathrm{AF}_{[\xi_{\mathrm{ref}};\xi(t)]}\,\mathrm{d}t \tag{5.1}$$

$$t_{\mathrm{ref}} = \int_0^{t_{\mathrm{BD}}} \mathrm{AF}_{[\xi_{\mathrm{ref}};\xi(t)]}\,\mathrm{d}t \tag{5.2}$$

$$t_{\mathrm{ref}} = \int_0^{t_{\mathrm{BD}}} \frac{\mathrm{TTF}(\xi_{\mathrm{ref}})}{\mathrm{TTF}(\xi(t))}\,\mathrm{d}t \tag{5.3}$$

Time-to-failure (TTF) is the respective failure time for the specified stress ξ and can be calculated with an applicable acceleration model of the investigated failure mechanism. It is used analogously to chapter 3.3 and refers to lifetimes under CVS. The *equivalent lifetime* under a reference CVS is denoted as t_{ref} and is substituted in the following sections with t_{CVS} to pronounce the distinct reference to constant stress. As already introduced, t_{BD} is the observed breakdown time of the conducted reliability measurement and is predominantly used as integration bound and afterwards substituted with t when a distribution formula is devised. On the contrary, calculated equivalent lifetimes under RVS are denoted as t_{RVS} when transforming reliability data back from constant stress conditions.

A characteristic quantity for ramp-stress tests is the ramp rate λ, which indicates the rate of stress level increase at every ramp time step:

$$\lambda = \frac{\mathrm{d}\xi}{\mathrm{d}t} \tag{5.4}$$

Assuming a linear ramp function, which is an approximation of the performed SSALT for sufficient small step sizes [Key17, JED03], the stress at time t can be calculated with:

$$\xi(t) = \lambda \cdot t \tag{5.5}$$

There are indications that the assumption of a linear ramp rate fails for too high ramp rates or too thick oxides due to impact ionization [Aal08]. In that case, the CE model and the methodology of the chapter 4 can be used for evaluation, fitting and extrapolation of RVS data with respect to the actual stress behavior of the step-stress levels and the stress transients located between them [Aal06, Aal07, Aal08].

The resulting ramp-stress CDF is plotted according to chapter 2.1.4 and is derived from the linearized Weibull distribution for probability plotting (cf. Eq. (2.11))

$$\mathrm{Wb_{RVS}} = \beta \cdot \ln\left(\frac{t_{\mathrm{ref}}}{t_{\mathrm{ref},\mathbf{63}}}\right) \tag{5.6}$$

in which t_{ref} from Eq. (5.3) can be plugged in.

Furthermore, a series of ramp-stress experiments with varying ramp rates λ can be used analogously to a series of CVS experiments with varying constant stress levels ξ for fitting acceleration model parameters [Ker07]. For this purpose, different sets of test parameters $(\xi_{\mathbf{63},i}, \lambda_i)$ for different ramp-stresses i are equated with another because of the assumption that with Eq. (5.3) different ramp-stress tests yield identical failure times $t_{\mathrm{ref},\mathbf{63}}$ for the same reference constant stress conditions. This equality can then be harnessed to derive a function $f\colon \xi_{\mathbf{63},i} \mapsto \lambda_i$ in such a way that the set of test parameters $(\xi_{\mathbf{63},i}, \lambda_i)$ can be used to fit the acceleration model parameters a and b analogous to the linearized acceleration models in Eq. (3.5).

In the following, the generic formula for RVS in Eq. (5.3) is elaborated together with its inverse functions, as well as the ramped CDF in Eq. (5.6), and the equations for parameter fitting for all four different TDDB voltage acceleration models in chapter 3.3. Thereby, experimental results of CVS and RVS measurements can be compared and correlated to one another. Additionally, these equations create the basis for parameter fitting with ramp-stress data, which can be another time-saving application of the reliability assessment with fWLR. Note that the variables a and b always relate to the respective individual acceleration model and are therefore ambiguously defined.

Since not only application-oriented but also the analytical investigation of ramp-stress methodology is of interest, the rearranged equations will be given in their analytical form where possible and likewise with and without the simplification of omitting the lower integration bound of Eq. (5.3) for all introduced acceleration models when appropriate. This differs essentially from the works of other authors in richness of detail and size of the scope and therefore represents original work on the topic of ramp-stress studies.

The Exponential *E* Model is defined as presented in Eq. (3.1) and is given as typical and linearized equation for the purpose of parameter fitting with CVS reliability data:

$$\mathrm{TTF} = a \cdot \exp\left(-b \cdot E\right) \tag{5.7}$$

$$\ln(\mathrm{TTF}) = (-b \cdot E) + \ln(a) \tag{5.8}$$

Continuing from Eq. (5.3), the equivalent reference lifetime $t_{\mathrm{CVS}} = t_{\mathrm{ref}}$ under CVS for the exponential E model in Eq. (5.7) is calculated as:

$$\begin{aligned} t_{\mathrm{CVS}} &= \int_0^{t_{\mathrm{BD}}} \frac{a \cdot \exp\left(-b \cdot E_{\mathrm{ref}}\right)}{a \cdot \exp\left(-b \cdot E(t)\right)} \, \mathrm{d}t \\ &= \int_0^{t_{\mathrm{BD}}} \exp\Big(b \cdot (\lambda t - E_{\mathrm{ref}})\Big) \, \mathrm{d}t \\ &= \left[\frac{1}{b \cdot \lambda} \cdot \exp\Big(b \cdot (\lambda t - E_{\mathrm{ref}})\Big)\right]_0^{t_{\mathrm{BD}}} \\ &= \frac{1}{b \cdot \lambda} \cdot \Big(\exp\big[b \cdot (\lambda t_{\mathrm{BD}} - E_{\mathrm{ref}})\big] - \exp\big[-b \cdot E_{\mathrm{ref}}\big]\Big) \end{aligned} \tag{5.9}$$

For pragmatic reasons, the second term $\exp\left[-b \cdot E_{\mathrm{ref}}\right]$ is usually considered to be negligibly small for values of E_{ref} [McP19], which are usually applied, and Eq. (5.9) can therefore be approximated by:

$$t_{\mathrm{CVS}} \approx \frac{1}{b \cdot \lambda} \cdot \Big(\exp\big[b \cdot (\lambda t_{\mathrm{BD}} - E_{\mathrm{ref}})\big]\Big) \tag{5.10}$$

The inverse functions of Eqq. (5.9) and (5.10) can be devised by solving the equations for $t_{\mathrm{BD}} = t_{\mathrm{RVS}}$:

$$t_{\mathrm{RVS}} = \frac{1}{b \cdot \lambda} \cdot \Big(\ln\big[1 + b \cdot \lambda t_{\mathrm{BD,CVS}} \cdot \exp\left(b \cdot E_{\mathrm{ref}}\right)\big]\Big) \tag{5.11}$$

$$\approx \frac{1}{b \cdot \lambda} \cdot \Big(\ln\big[b \cdot \lambda t_{\mathrm{BD,CVS}}\big] + (b \cdot E_{\mathrm{ref}})\Big) \tag{5.12}$$

With these functions, CVS failure data can be transformed and compared to RVS data for model verification and reliability analysis.

The resulting CDF of a voltage ramp test with applied exponential E model can be calculated with Eqq. (5.6) and (5.10) and is given by

$$\begin{aligned} \mathrm{Wb}_{\mathrm{RVS}} &\approx \beta \cdot \Big(b\left(E_{\mathrm{BD}} - E_{\mathrm{BD},\mathbf{63}}\right)\Big) \\ &\approx \beta \cdot \Big(b \cdot \lambda\left(t_{\mathrm{BD}} - t_{\mathrm{BD},\mathbf{63}}\right)\Big) \end{aligned} \tag{5.13}$$

which does not satisfy the form of a linearized Weibull distribution, resulting in a bend line in the Weibull probability plot due to the logarithmic time-axis.

The ramp rate λ for a linear increasing voltage ramp can be set in relation to the electrical breakdown field $E_{\mathrm{BD},\mathbf{63}}$ for a ramp-stress measurement series by equating, rearranging and linearizing Eq. (5.9) for different sets of measurement parameters $(E_{\mathbf{63},i}, \lambda_i)$

$$\lambda = \frac{1}{a'} \cdot \exp\left(b \cdot E_{\mathrm{BD},\mathbf{63}}\right)$$

$$\ln(\lambda) = b \cdot E_{\mathrm{BD},\mathbf{63}} - \ln(a') \tag{5.14}$$

and subsequently used for parameter fitting of a' and b with RVS failure data. Whereas b is the universal stress acceleration parameter for ramp and constant voltage stress, the constant undefined intercept of this linear equation is by convention denoted as a' to clarify that a' is the respective parameter of RVS as a is to CVS.

The Exponential $\sqrt{E}$ Model is given as typical and linearized equation as already presented in Eq. (3.2):

$$\mathrm{TTF} = a \cdot \exp\left(-b \cdot \sqrt{E}\right) \tag{5.15}$$

$$\ln(\mathrm{TTF}) = \left(-b \cdot \sqrt{E}\right) + \ln(a) \tag{5.16}$$

From Eq. (5.15), the equivalent lifetime under reference CVS stress is then calculated with Eq. (5.3) as

$$\begin{aligned}
t_{\mathrm{CVS}} &= \int_0^{t_{\mathrm{BD}}} \frac{a \cdot \exp\left(-b \cdot \sqrt{E_{\mathrm{ref}}}\right)}{a \cdot \exp\left(-b \cdot \sqrt{E(t)}\right)} \,\mathrm{d}t \\
&= \int_0^{t_{\mathrm{BD}}} \exp\left[b \cdot \left(\sqrt{\lambda t} - \sqrt{E_{\mathrm{ref}}}\right)\right] \mathrm{d}t \\
&= \left[\frac{2\left(b \cdot \sqrt{\lambda t} - 1\right)}{\lambda \cdot b^2} \cdot \exp\left[b \cdot \left(\sqrt{\lambda t} - \sqrt{E_{\mathrm{ref}}}\right)\right]\right]_0^{t_{\mathrm{BD}}} \\
&= \left\{\frac{2\left(b \cdot \sqrt{\lambda t_{\mathrm{BD}}} - 1\right)}{\lambda \cdot b^2} \cdot \exp\left[b \cdot \left(\sqrt{\lambda t_{\mathrm{BD}}} - \sqrt{E_{\mathrm{ref}}}\right)\right]\right\} \\
&\quad - \left\{\frac{-2}{\lambda \cdot b^2} \cdot \exp\left[b \cdot \left(-\sqrt{E_{\mathrm{ref}}}\right)\right]\right\}
\end{aligned} \tag{5.17}$$

which results in a fairly good approximation by omitting the lower integration bound:

$$t_{\mathrm{CVS}} \approx \left\{\frac{2\left(b \cdot \sqrt{\lambda t_{\mathrm{BD}}} - 1\right)}{\lambda \cdot b^2} \cdot \exp\left[b \cdot \left(\sqrt{\lambda t_{\mathrm{BD}}} - \sqrt{E_{\mathrm{ref}}}\right)\right]\right\} \tag{5.18}$$

When solving Eqq. (5.17) and (5.18) for $t_{\mathrm{BD}} = t_{\mathrm{RVS}}$, one obtains respectively

$$t_{\mathrm{RVS}} = \frac{1}{b^2\lambda}\left\{1 + W\left[\frac{-2 + b^2 \cdot \exp\left(b \cdot \sqrt{E_{\mathrm{ref}}}\right)\lambda t_{\mathrm{BD,CVS}}}{2\,\mathrm{e}}\right]\right\}^2 \tag{5.19}$$

$$t_{\mathrm{RVS}} \approx \frac{1}{b^2\lambda}\left\{1 + W\left[\frac{b^2}{2}\exp\left(b \cdot \sqrt{E_{\mathrm{ref}}} - 1\right)\lambda t_{\mathrm{BD,cvs}}\right]\right\}^2 \tag{5.20}$$

with $W(x)$ being the the Lambert W function in MATLAB [Mat20] or the product logarithm function in Mathematica [Wol20], which are both defined as the solution y of the equation $x = y\exp(y)$.

Using the simplification from Eq. (5.18), the ramped CDF in a Weibull probability plot can be expressed as

$$\begin{aligned}\mathrm{Wb}_{\mathrm{RVS}} &\approx \beta \cdot \ln\left[\frac{b \cdot \sqrt{E_{\mathrm{BD}}} - 1}{b \cdot \sqrt{E_{\mathrm{BD},\mathbf{63}}} - 1}\right] + \beta \cdot \left[b \cdot \left(\sqrt{E_{\mathrm{BD}}} - \sqrt{E_{\mathrm{BD},\mathbf{63}}}\right)\right] \\ &\approx \beta \cdot \ln\left[\frac{b \cdot \sqrt{\lambda t_{\mathrm{BD}}} - 1}{b \cdot \sqrt{\lambda t_{\mathrm{BD},\mathbf{63}}} - 1}\right] + \beta \cdot \left[b \cdot \sqrt{\lambda}\left(\sqrt{t_{\mathrm{BD}}} - \sqrt{t_{\mathrm{BD},\mathbf{63}}}\right)\right]\end{aligned} \tag{5.21}$$

which is neither an intuitively accessible formula nor does it have a linear representation in a Weibull probability plot.

For the simplified version in Eq. (5.18), we can also derive this expression

$$\begin{aligned}\lambda &\approx \frac{1}{a'} \cdot \left(b \cdot \sqrt{E_{\mathrm{BD},\mathbf{63}}} - 1\right) \cdot \exp\left(b \cdot \sqrt{E_{\mathrm{BD},\mathbf{63}}}\right) \\ \ln(\lambda) &\approx \ln\left(b \cdot \sqrt{E_{\mathrm{BD},\mathbf{63}}} - 1\right) + \left(b \cdot \sqrt{E_{\mathrm{BD},\mathbf{63}}}\right) - \ln(a')\end{aligned} \tag{5.22}$$

which cannot be further simplified or rearranged to be utilized for parameter fitting in an ordinary linear plot type. Therefore, a numerical non-linear fitting algorithms with method of least squares is used to derive the model parameters for the exponential $\sqrt{E}$ model.

The Power-Law *U* Model does not depend on the electrical field E but instead on the electrical voltage U and is defined in Eq. (3.3) as:

$$\mathrm{TTF} = a \cdot U^{-b} \tag{5.23}$$

$$\ln(\mathrm{TTF}) = (-b) \cdot \ln(U) + \ln(a) \tag{5.24}$$

By using Eqq. (5.3) and (5.23), the RVS lifetimes t_{BD} can be transformed into CVS lifetimes $t_{\text{CVS}} = t_{\text{ref}}$:

$$\begin{aligned}
t_{\text{CVS}} &= \int_0^{t_{\text{BD}}} \frac{a \cdot U_{\text{ref}}^{-b}}{a \cdot (U(t))^{-b}} \, \mathrm{d}t \\
&= \int_0^{t_{\text{BD}}} \left(\frac{U_{\text{ref}}}{\lambda t}\right)^{-b} \mathrm{d}t \\
&= \left[\frac{1}{b+1} \cdot \left(t \cdot \left(\frac{U_{\text{ref}}}{\lambda t}\right)^{-b}\right)\right]_0^{t_{\text{BD}}} \\
&= \frac{1}{b+1} \cdot \left(t_{\text{BD}} \cdot \left(\frac{U_{\text{ref}}}{\lambda t_{\text{BD}}}\right)^{-b}\right)
\end{aligned} \tag{5.25}$$

The inverse function of Eq. (5.25) for $t_{\text{BD}} = t_{\text{RVS}}$ is devised as:

$$t_{\text{RVS}} = \sqrt[b+1]{t_{\text{BD,CVS}} \cdot (b+1) \cdot U_{\text{ref}}^{b} \cdot \lambda^{-b}} \tag{5.26}$$

Inserting the obtained Eq. (5.25) for $t_{\text{CVS}} = t_{\text{ref}}$ into Eq. (5.6) results in the CDF of a voltage ramp test with applied power-law U model

$$\begin{aligned}
\text{Wb}_{\text{RVS}} &= \beta\,(b+1) \cdot \ln\left(\frac{U_{\text{BD}}}{U_{\text{BD},\mathbf{63}}}\right) \\
&= \beta\,(b+1) \cdot \ln\left(\frac{t_{\text{BD}}}{t_{\text{BD},\mathbf{63}}}\right)
\end{aligned} \tag{5.27}$$

which takes the form of a linearized Weibull distribution (cf. Eq. (2.11)). Therefore in a Weibull probability plot, reliability data of a RVS test are plotted in a straight line with a slope of $\beta(b+1)$ for this particular acceleration model. Moreover, this relation of the Weibull slopes of measured CVS and RVS can also be used for parameter fitting.

Further parameter fitting of the power-law U model can be performed with a linearized relation of the ramp rate λ and the characteristic breakdown voltage $U_{\text{BD},\mathbf{63}}$ of a RVS measurement series:

$$\begin{aligned}
\lambda &= \frac{1}{a'} \cdot U_{\text{BD},\mathbf{63}}^{b+1} \\
\ln(\lambda) &= (b+1) \cdot \ln(U_{\text{BD},\mathbf{63}}) - \ln(a')
\end{aligned} \tag{5.28}$$

The Exponential 1/E Model is the acceleration model of Eq. (3.4) and is defined as the following:

$$\mathrm{TTF} = a \cdot \exp\left(\frac{b}{E}\right) \tag{5.29}$$

$$\ln(\mathrm{TTF}) = \frac{b}{E} + \ln(a) \tag{5.30}$$

By combining Eqq. (5.3) and (5.29), an equation for t_{CVS} can be obtained

$$\begin{aligned} t_{\mathrm{CVS}} &= \int_0^{t_{\mathrm{BD}}} \frac{a \cdot \exp\left(\frac{b}{E_{\mathrm{ref}}}\right)}{a \cdot \exp\left(\frac{b}{E(t)}\right)} \mathrm{d}t \\ &= \int_0^{t_{\mathrm{BD}}} \exp\left(b \cdot \left(\frac{1}{E_{\mathrm{ref}}} - \frac{1}{\lambda t}\right)\right) \mathrm{d}t \\ &= \left[\frac{1}{\lambda} \cdot \exp\left(\frac{b}{E_{\mathrm{ref}}}\right) \cdot \left(b \cdot \mathrm{Ei}\left[-\frac{b}{\lambda t}\right] + \lambda t \cdot \exp\left(-\frac{b}{\lambda t}\right)\right)\right]_0^{t_{\mathrm{BD}}} \\ &= \frac{1}{\lambda} \cdot \exp\left(\frac{b}{E_{\mathrm{ref}}}\right) \cdot \left(b \cdot \mathrm{Ei}\left[-\frac{b}{\lambda t_{\mathrm{BD}}}\right] + \lambda t_{\mathrm{BD}} \cdot \exp\left(-\frac{b}{\lambda t_{\mathrm{BD}}}\right)\right) \end{aligned} \tag{5.31}$$

with $\mathrm{Ei}(t)$ being the exponential integral defined as $\mathrm{Ei}(t) = \int_{-\infty}^{x} \frac{e^t}{t} \mathrm{d}t$ [Bro12, p. 526, Eq. (8.102a)]. As the lower integration bound equals 0, this equation cannot be further simplified like the exponential E and exponential $\sqrt{E}$ models.

For t_{RVS} only a numerical solution can be calculated because of the difficulty to rearrange the equation and solve for t_{BD}, which is given inside as well as outside of the inverse Ei function.

Consequently for similar reasons, inserting Eq. (5.31) into Eq. (5.6) does not yield a term for the RVS CDF that could be rearranged and simplified enough to give an meaningful formula for probability plotting:

$$\begin{aligned} \mathrm{Wb}_{\mathrm{RVS}} &= \beta \cdot \ln\left(\frac{b \cdot \mathrm{Ei}\left[-\frac{b}{E_{\mathrm{BD}}}\right] + E_{\mathrm{BD}} \cdot \exp\left(-\frac{b}{E_{\mathrm{BD}}}\right)}{b \cdot \mathrm{Ei}\left[-\frac{b}{E_{\mathrm{BD,63}}}\right] + E_{\mathrm{BD,63}} \cdot \exp\left(-\frac{b}{E_{\mathrm{BD,63}}}\right)}\right) \\ &= \beta \cdot \ln\left(\frac{b \cdot \mathrm{Ei}\left[-\frac{b}{\lambda t_{\mathrm{BD}}}\right] + \lambda t_{\mathrm{BD}} \cdot \exp\left(-\frac{b}{\lambda t_{\mathrm{BD}}}\right)}{b \cdot \mathrm{Ei}\left[-\frac{b}{\lambda t_{\mathrm{BD,63}}}\right] + \lambda t_{\mathrm{BD,63}} \cdot \exp\left(-\frac{b}{\lambda t_{\mathrm{BD,63}}}\right)}\right) \end{aligned} \tag{5.32}$$

Similar to Eq. (5.22) of the exponential $\sqrt{E}$ model, the relation between the ramp rate λ and $E_{\mathrm{BD,63}}$ for a RVS measurement series cannot be linearized for a parameter fitting plot but instead be used for numerical non-linear least-squares parameter fitting:

$$\lambda = \frac{1}{a'} \cdot \left(b \cdot \mathrm{Ei}\left[-\frac{b}{E_{\mathrm{BD,63}}} \right] + E_{\mathrm{BD,63}} \cdot \exp\left(-\frac{b}{E_{\mathrm{BD,63}}} \right) \right)$$

$$\ln(\lambda) = \ln \left(b \cdot \mathrm{Ei}\left[-\frac{b}{E_{\mathrm{BD,63}}} \right] + E_{\mathrm{BD,63}} \cdot \exp\left(-\frac{b}{E_{\mathrm{BD,63}}} \right) \right) - \ln(a') \qquad (5.33)$$

Concluding for the TDDB Acceleration Models the model acceleration factors between constant and ramp-stress Weibull failure distributions $\mathrm{AF} = t_{\mathrm{CVS}}/t_{\mathrm{RVS}}$ at different times $t_{\mathrm{RVS}} = t_{\mathrm{BD}}$ are given in Fig. 5.3. Additionally to the model AFs, the empirical acceleration factor on a point-to-point basis of two measured failure distributions with the same number of fails as well as of the fitted Weibull distributions is given for comparison. Note, that while the AF of two measured data points is model-independent, calculating the ratio of the fitted Weibull CDFs of these data already implies the power-law U model as will be discussed later in this section. This is the reason for the similar behavior of the empirical and power-law U model AF over the wide range of the plot. A logarithmic x-axis would further highlight this characteristic, as both AF behaviors would become straight lines in Fig. 5.3.

The obtained AFs are also present when used for the subsequent transformation of RVS and CVS failure data in Weibull probability plots as illustrated in Fig. 5.4. Due to the very different slopes of the distributions on the logarithmic time-axis in the Weibull probability plots, the AF variations between acceleration models are scaled differently in Figs. 5.4a and 5.4b and are therefore perceived as differently distinct to the eye. Starting from the CVS Weibull CDF in Fig. 5.4a, the calculated RVS data of the four stress acceleration models can hardly be distinguished over a wide Weibit range, even at full zoom, and especially for experimental data that are subject to statistical variations. However as mentioned before, only the power-law U model exhibits Weibull-like RVS distributions (see Eq. (5.27)). Thus, when transforming ramp failure data to CVS data (see Fig. 5.4b) using a fitted Weibull RVS CDF is predestined to yield a straight Weibull CDF only for the power-law U model and a bent behavior for the other stress acceleration models. This is a source of confusion, as only the power-law U model seems to produce valid effective CVS failure times that are Weibull distributed. The mistake here is the erroneous assumption of treating the RVS data as intrinsically Weibull-like.

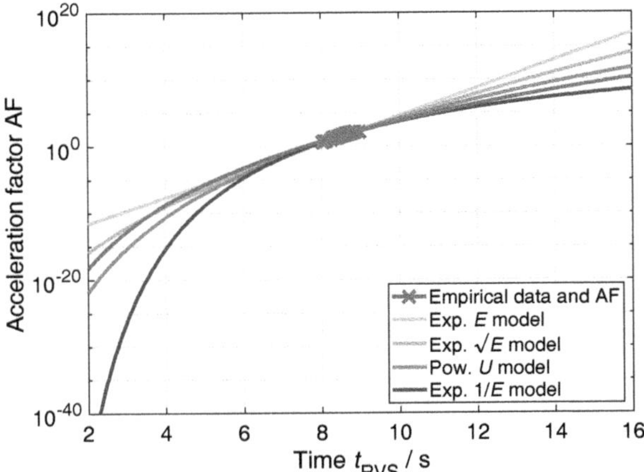

Figure 5.3: Empirical and model acceleration factors of the CVS and RVS failure times displayed in Fig. 5.4. Blue crosses depict the AF between two failure times with the same cumulative fraction failed. The lifetime ratio between the fitted Weibull CDFs of the failure data is illustrated with a blue line.

Having said this, transforming RVS data into CVS data with an unknown acceleration model will result in a slightly bent behavior for all but the appropriate acceleration model. This property could be used as an approach to determine the corresponding acceleration model for a set of experimental RVS data. This is due to the fact that the nature of stress acceleration is already intrinsic to the RVS data, whereas CVS data need to be accelerated to other stress levels before assessing different stress acceleration models is possible. However, these RVS data would have to be large enough in sample size and not be affected by any other influence that might result in a bending of the CDF tails, like emerging extrinsic failure behavior, or inappropriate yield-censoring of the experimental failure data, in order to be suitable for model evaluation.

One might be tempted to use Poisson area scaling (see Eq. (3.8)) for the RVS data to artificially extend the empirically accessible Weibit range [Ker06a, Ker07, Rah11], but this proposition would again assume and preselect the power-law U model over the other acceleration models and lead to a biased model verification.

Summarizing this section, the common practice of plotting and fitting any ramp-stress failure distribution usually as a Weibull distribution [McP19] is acceptable for practical aspects but is only correct for the power-law U model from a stricter mathematical point of view and can lead to bias and errors in the model evaluation if it is not adequately taken into account.

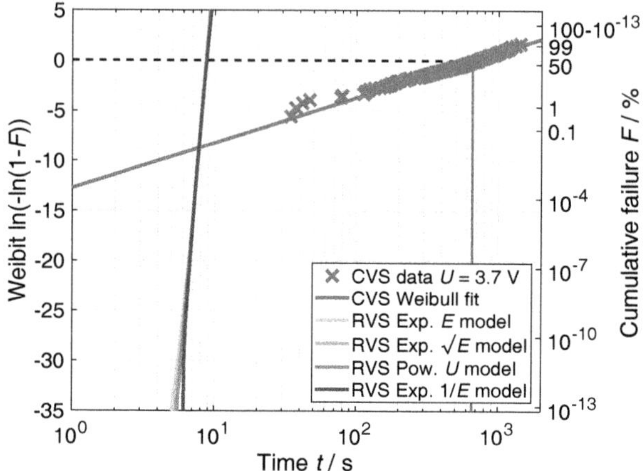

(a) Transformation of the depicted Weibull CDF fit of experimental CVS failure data to RVS CDFs with the ramp characteristic of the failure data in subfigure (b).

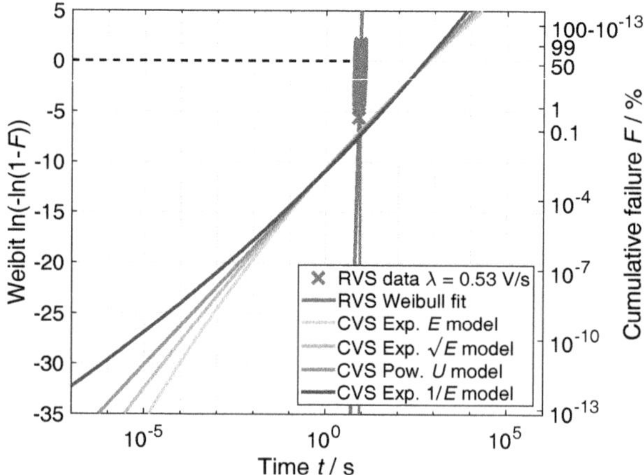

(b) Transformation of the depicted Weibull CDF fit of experimental RVS failure data to CVS CDFs with the stress voltage of the failure data in subfigure (a).

Figure 5.4: Transformation of CVS and RVS failure data according to the applied TDDB acceleration models for a particular wide Weibit range.

5.3 Validation of Stress-Equivalence

After analysis of the specific differences of the four acceleration models and their effects on the transformation of CVS into RVS data and vice versa, the key element of this section is to verify the equivalence of reliability data obtained by RVS and CVS measurements and discuss these findings in the context of other works in literature.

In order to be able to make use of RVS measurements and fWLR methodology for reliability statements during or after technology or product development or production, it has to be verified that the data obtained from RVS and CVS measurements are equivalent in terms of an equal amount of endured stress and comparable occurrence of damage and failure of the DUTs.

For that reason, failure data of ramp-stress measurements are transformed into effective CVS failure times with different constant stress voltages according to section 5.2 and compared with generic CVS measurements of those stresses in Fig. 5.5. It can be observed that, although the Weibull slope of the generic CVS distributions is subject to statistical variation, the slope of the transformed RVS data is unaltered for different constant voltage stresses. This is not surprising as analogous to the power-law U model, the Weibull slope of transformed CDFs depends only on the steepness of the initial CDF and the stress acceleration parameter b of the respective model.

This leads to the conclusion that non-matching distribution shape parameters of transformed and generic reliability data are an indication of either an inappropriately estimated stress acceleration parameter b or an altered failure mechanism due to the performed stressor ramping. This might be caused by the stress ramp exceeding the maximum validated stress level range of the acceleration model and giving rise to new or altered failure mechanisms or by accumulating stresses form domains of various failure mechanism characteristics, which has to be studiously avoided (see Fig. 2.15). Likewise, if CVS data show a declining or increasing trend for the Weibull slope over a wider stress range, which is also an indication of an altered failure mechanism, the β of transformed RVS data is likely to better fit those measurements with similar stress-to-failure voltages than deviating ones.

Additionally, the different Weibull slopes between the individual acceleration models of the transformed RVS distribution depicted in Fig. 5.5 can be attributed to the model behavior in the observed Weibit range in Fig. 5.4b. Hence, the exponential E model will exhibit the lowest Weibull slope after transformation to a CVS distribution, followed by the exponential $\sqrt{E}$ model, and the power-law U model, with the exponential $1/E$ model featuring the steepest Weibull slope of these four acceleration models.

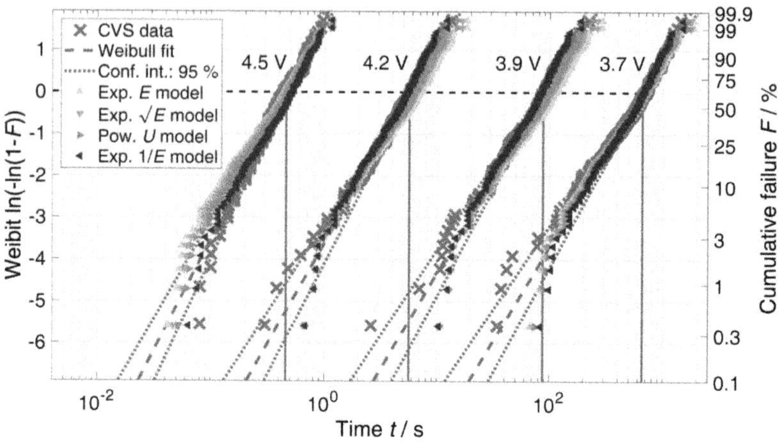

Figure 5.5: Transformation of RVS failure data with $\lambda = 0.26\,\mathrm{V/s}$ to CVS for all four different TDDB acceleration models onto existing CVS failure distributions with different stress voltages.

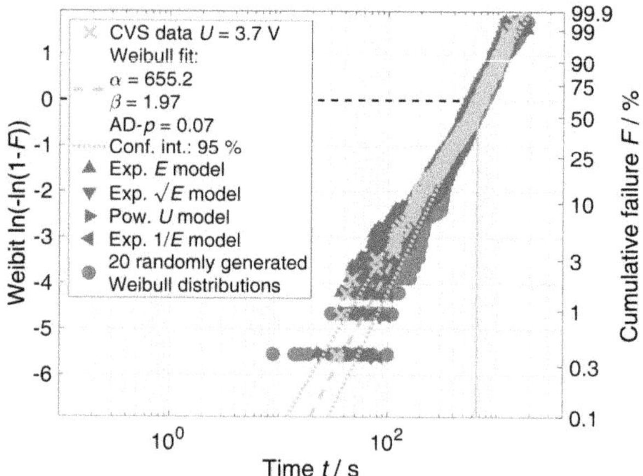

Figure 5.6: Transformation of RVS failure data with five different ramp rates of all four acceleration models onto the same CVS data with stress voltage $U = 3.7\,\mathrm{V}$ with confidence intervals and 24 artificially generated Weibull CVS distributions for comparison with the statistical variance of the transformed RVS data.

Focusing on the other Weibull parameter, the characteristic lifetime α or t_{63}, a connection between the deviation of the t_{63} of the transformed RVS data and the generic CVS measurements and the extent to which the respective acceleration model can fit the measured TTF–stress relation (see Fig. 3.13) is observed.

This connection has also been investigated in [Wu09a] and it is demonstrated that transformed RVS data to different constant stress levels can only reproduce the measured CVS data to a certain degree in specific stress-level regions similar to Fig. 5.5.

In order to better assess the extent of the variation between the acceleration models when it comes to reproduce generic CVS data, several RVS measurements are transformed and overlaid with the CVS distribution of the same stress voltage with all four acceleration models in Fig. 5.6 and compared to the statistical variance of artificial Weibull data that are generated randomly with MATLAB. This comparison yields that transformed RVS distributions exhibit the same statistical variation to the CVS data as randomly generated data and are therefore suited to reproduce failure behavior of CVS measurements.

From the data shown in Fig. 5.5, a single best-fitting acceleration model could not be identified for the investigated stress range, as the variation of β between the CVS measurements is comparable to that between transformed distributions of the individual acceleration models. Additionally, t_{63} is matched comparably accurate as this is the stress range to which all four acceleration models are fitted to match these failure distributions.

Unlike this course of action, works by other authors often compare the results transformed with only one assumed acceleration model to measured CVS data. Nevertheless, they also consistently show the equivalence of failure behavior of constant and ramp-stress measurements if data are collected thoroughly and with cautious methodology [McP85, Wol85, Ros96, Ker06b, Ker07]. This includes a not too high ramp-rate in order to not reach stress-levels that can alter the failure mechanism and a determined range of validity for which the CVS measurements are obtained.

On the topic of stress equivalence, these findings imply that for the given measurement and stress-level range, the transformed RVS data are as equivalent to the generic CVS measurements as any remeasured CVS data are expected to be. Therefore, reliability statements that are obtained with CVS data can in principal also be derived from RVS data with less experimental time expenses, which is the primal objective of fWLR.

5.4 Model Fitting with Ramp Tests

After verifying the stress equivalence of ramp and constant stress tests, the methodology of RVS can hence be extended for the additional application in model parameter fitting. This would further reduce the amount of time needed for reliability qualification and fit well within the fWLR toolbox, since additional time intensive CVS tests would otherwise be necessary to obtain model parameters or the same statistical accuracy.

Analogously to the parameter fitting of CVS failure data for which linearized time-to-failure–stress diagrams (see Fig. 3.11) are utilized, parameter fitting for RVS reliability data can be accomplished by employing the λ–stress-to-failure dependencies in Eqq. (5.14) and (5.28) in the respective linearized diagrams of Fig. 5.7. However, Eqq. (5.22) and (5.33) cannot be linearized for similar plots and are instead fitted using a numerical non-linear curve fitting algorithm with a comparable least square approach. As a consequence of the deliberate choice of $\ln(\lambda)$ as the y-axis for all acceleration models, the resulting GOF metrics are directly comparable to those in Fig. 5.7.

The obtained fitted models for this technique are plotted in Fig. 5.8 for all four TDDB acceleration models and compared with the results of the parameter fitting with CVS data in Table 5.1. Although the parameters a for CVS and a' for RVS are obtained similarly as a constant offset of the linear dependency, they do not represent equivalent quantities and are therefore not strictly comparable. However, the universal stress-acceleration parameter b meets this property and gives similar estimates $\hat{b}$ for both parameter fitting techniques with overlapping confidence intervals of 95 %. The estimated values of $\hat{b}$ obtained from RVS data are in this case slightly smaller than those from CVS but can be considered equivalent in respect to statistical variation. Hence, model data obtained by fWLR from RVS testing have proven to yield the same results than parameter fitting with CVS data and can be used for elaborated reliability studies.

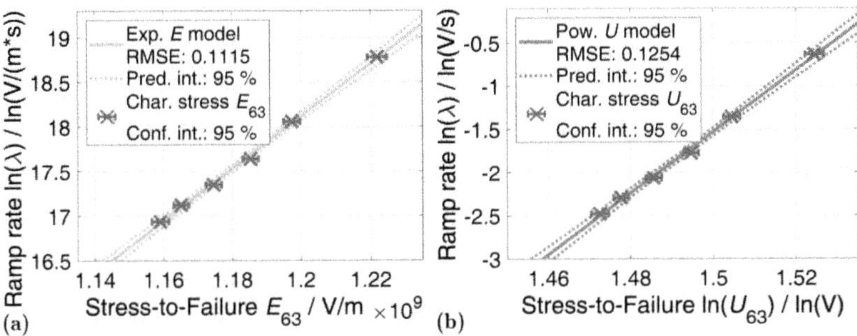

Figure 5.7: Parameter fit for the **(a)** exponential E model from linearized λ–E_{63} plot (see Eq. (5.14)) and **(b)** power-law U model from linearized λ–U_{63} plot (see Eq. (5.28)). The root mean squared error (RMSE) value of the exponential $\sqrt{E}$ model equals 0.1185, that of the exponential $1/E$ model equals 0.1256 for comparison.

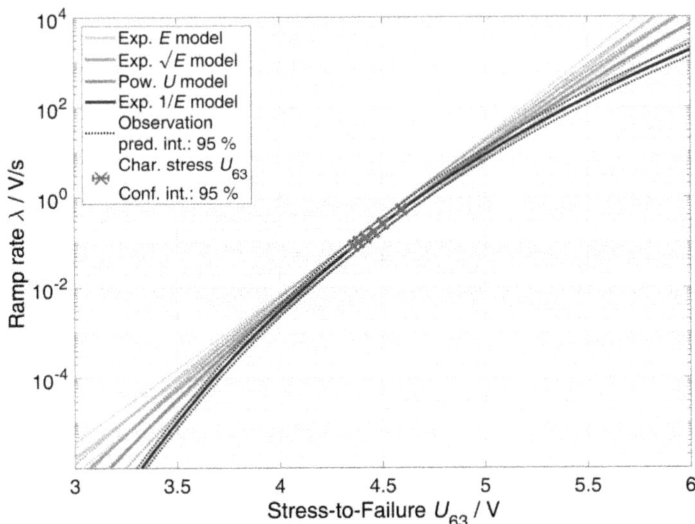

Figure 5.8: All four fitted TDDB acceleration models in a λ–U_{63} plot with observation prediction intervals (cf. Fig. 3.13b).

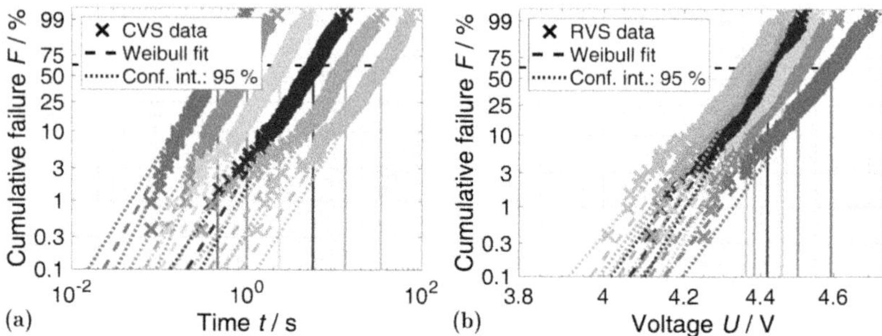

Figure 5.9: Experimental Weibull data for model fitting: **(a)** CVS data CDFs with constant voltages U from 4.0 V (light blue) to 4.5 V (dark blue) in steps of 0.1 V and an average Weibull slope $\langle\beta_{\mathrm{CVS}}\rangle = 2.17 \pm 0.10$ and **(b)** RVS data CDFs with ramp rates λ from 0.08 V/s (light blue) to 0.53 V/s (dark blue) and an average Weibull slope $\langle\beta_{\mathrm{RVS}}\rangle = 78 \pm 3$ (see Fig. 5.2a).

Table 5.1: Acceleration model fit parameters obtained from different techniques given in their respective units with the dimensions s, V, and m.

Model:	Exp. E fit	Exp. $\sqrt{E}$ fit	Pow. U fit	Exp. $1/E$ fit
CVS analysis and parameter fitting (see Fig. 5.9a, cf. Fig. 3.13)				
$\hat{a}$	$\exp(36 \pm 2)$	$\exp(72 \pm 3)$	$\exp(54 \pm 1)$	$\exp(-35 \mp 1)$
$\hat{b}$	$(3.1 \pm 0.1)\cdot 10^{-8}$	$(2.1 \pm 0.1)\cdot 10^{-3}$	$(3.6 \pm 0.1)\cdot 10^{1}$	$(4.1 \pm 0.1)\cdot 10^{10}$
RVS analysis and parameter fitting (see Figs. 5.7 and 5.8)				
$\hat{a'}$	$\exp(17 \pm 2)$	$\exp(55 \pm 5)$	$\exp(55 \pm 4)$	$\exp(-33)$
$\hat{b}$	$(2.9 \pm 0.2)\cdot 10^{-8}$	$(2.0 \pm 0.1)\cdot 10^{-3}$	$(3.5 \pm 0.1)\cdot 10^{1}$	$3.9\cdot 10^{10}$
RVS and CVS Weibull distribution slope fitting (see Figs. 5.9a and 5.9b)				
$\hat{b}$	—	—	$(3.5 \pm 0.2)\cdot 10^{1}$	—

Another advantageous relation for parameter fitting can be utilized in particular for the power-law U model, as the fails of CVS and RVS tests are both Weibull distributed, given in Eq. (2.11). With this, Eq. (5.27) can be used to establish an expression for $\hat{b}$

$$\mathrm{Wb}_{\mathrm{RVS}} = \underbrace{\beta_{\mathrm{CVS}}\,(b+1)}_{\text{slope: }\beta_{\mathrm{RVS}}} \cdot \ln\left(\frac{t}{t_{63}}\right)$$

$$\Rightarrow \quad \hat{b} = \frac{\langle\beta_{\mathrm{RVS}}\rangle}{\langle\beta_{\mathrm{CVS}}\rangle} - 1 \tag{5.34}$$

where $\langle\beta_{\mathrm{CVS}}\rangle$ and $\langle\beta_{\mathrm{RVS}}\rangle$ are the respective means of the Weibull slope or shape parameter β in Fig. 5.9 (cf. [Ker07, Fig. 8]). The resulting value given in Table 5.1 does correspond well to the $\hat{b}$ value obtained either from CVS or RVS measurements. The result from just a single constant and ramp-stress measurement might deviate by a larger amount. Also note that in contrast to the formerly described methods which use the scale parameter α, i.e. time-to-failure or stress-to-failure, respectively, for fitting purposes, this model fitting technique solely uses the Weibull shape parameter β. Due to the cumulative nature of CDFs, the scale parameter is generally assumed to be more stable than the shape parameter in terms of variation [Str09]. Therefore, fitting methods based on the shape parameter β are generally expected to yield wider confidence intervals for the estimated acceleration parameters than those based on α, despite using double the measurement data. These findings are reflected in the values stated in Table 5.1.

As presented, different model parameter fitting techniques for RVS data perform equivalently well as the standard CVS model parameter fitting stated in section 3.3. Consequently, faster and more targeted ramp-stress tests in fWLR are very suitable to assist in reliability assessment and monitoring during technology or product development.

6 Conclusion and Outlook

In this work significant progress has been made on pressing issues of today's reliability qualification processes in the automotive industry for original equipment manufacturers and the associated semiconductor supply-chain. With the necessary usage of leading-edge semiconductor technologies in applications like driver assistance systems, autonomous driving, electric mobility and extended car connectivity, existing stress test qualification plans with fixed standardized test conditions are not sufficient to secure the typical high reliability standards required in the automotive industry. While new industry-wide qualification methods based on mission profile stress requirements are necessary and already in use, a physical justification as well as the mathematical foundation for transforming mission profile stresses into test conditions has finally been provided and empirically verified for the first time.

For this topic of mission profile stress transformation, cumulative damage models play an essential part. Three of the most accepted models, i.e. the cumulative exposure (CE), the tampered random variable (TRV), and the tampered failure rate (TFR) model, have been introduced and investigated for their applicability on established and state-of-the-art semiconductor technologies. On the exemplary failure mechanism time-dependent dielectric breakdown (TDDB) a concept of transforming mission profiles and alternating step-stress accelerated life testing (SSALT) stress sequences into equivalent effective stress levels or effective stress times for the two stressors of TDDB, voltage and temperature, was successfully demonstrated and empirically verified on 22 nm high-k metal-gate technology from GLOBALFOUNDRIES and university metal–oxide–semiconductor (MOS) capacitors. Additionally, this transformation was shown to be independent of the underlying acceleration model as it can be used exclusively with empirically determined acceleration factors. This greatly improves the applicability and transferability to other failure mechanisms and acceleration models.

Three essential properties of cumulative damage models necessary for the presented mission profile transformation were identified: Firstly, it has to be physically correct to divide a stressor into segments. Secondly, the sequence of segmented stress levels can be changed without any alteration of the failure physics in order to transform a stress time-series into a stress histogram. Finally, there has to be a physically justified and empirically verified method for the cumulative damage model to obtain an effective stress from a given stress histogram. For this method of transforming stress histograms into effective stress levels or stress times, it was shown for cumulative damage models, which are commutative in regards to the stress sequence like the CE and TRV model, that the acceleration factor of an equivalent constant effective stress distribution of cyclical SSALT can be derived by the weighted harmonic mean of the acceleration factors for the individual step-stress levels.

The first reported alternating SSALT measurements have been conducted on semiconductor devices for the failure mechanism TDDB for both stressors voltage and temperature. With that, the applicability of the CE and TRV model has been verified for this utilization, as the failure distributions are in agreement to the beforehand simulated failure behaviors and converged towards the calculated and measured effective stress distribution. In contrast, the TFR model was rejected for TDDB on the investigated devices for cumulative voltage stress. Furthermore, for Weibull shape parameters differing from 1, it was demonstrated that the TFR model exhibits deviating behavior for different permutations of stress level sequences and that therefore the stress sequence is of high importance for any reliability predictions conducted with the TFR model.

Moreover, the presented concept of transforming mission profiles and stress histograms into effective stress levels or stress times was extended onto multi-dimensional mission profiles with combined stressors that exhibited interdependencies. For the first time, cyclical SSALT was performed with interdependent temperature and voltage stresses for TDDB and revealed the significant impact of stressor interdependencies on reliability predictions. This was additionally emphasized by the comparison of three real-world mission profiles for automotive applications and demonstrated the consequences of unintentionally neglecting stressor interdependencies for the reliability qualification process.

In addition, the applicability of the CE and TRV model for TDDB was expanded from step-stress to ramp-stress testing. The main purpose of ramp-stress tests in fast wafer level reliability (fWLR) is to gather reliability data in shorter time and with better control over test durations and thus in larger numbers and with better statistics with ramp voltage stress (RVS) tests than with conventional constant voltage stress (CVS) tests. Within this scope, considerable progress has been made to further extend the areas of application for RVS testing.

On that account, a uniquely comprehensive overview and extensive analysis of the four major TDDB acceleration models exponential E, exponential $\sqrt{E}$, power-law U, and exponential $1/E$ model was presented for the investigation and transformation of RVS failure times and reliability distributions into CVS and vice versa. The common misconception of all RVS distributions being Weibull-like was resolved and instead, a method of using RVS distributions and the transformation into CVS data for model verification purposes was clarified and outlined.

Furthermore in accordance to the existing results from literature, the conducted measurements demonstrated stress equivalence of RVS and CVS failure data and reliability statements obtained by them. Nonetheless, the understudied properties and implications of different acceleration models on the Weibull slope and scale parameters of transformed RVS data have been elaborated for the detection of inappropriately fitted acceleration models and altered failure mechanisms.

Looking forward the herein developed and presented concept of transforming mission profile stresses into effective constant stress levels is based on the applicability of the CE model on the failure mechanism TDDB for the investigated semiconductor technologies. In order to apply this concept of reliability assessment comprehensively for the semiconductor and automotive industry, further research is needed to verify respective cumulative damage models likewise for other known failure mechanisms and existing semiconductor technologies. For the investigation of cumulative damage behavior, the demonstrated method of cyclical stress testing (CST) is recommended for the reason that repeated stress level alternations will emphasize the marginal differences between cumulative damage models.

Additionally, the properties of the CE and TRV model for stress sequences with differing Weibull slopes, indicating altered failure mechanisms, and on devices with self-healing capacities would considerably advance the understanding and model building of cumulative damage behavior.

On the subject of ramp-stress testing, the characteristics of area scaling for RVS distributions with respect to the individual acceleration models remain an open topic for further analysis, as Poisson area scaling is expected to not apply in most cases. Furthermore, focused studies on the functionality and feasibility of model verification and rejection on basis of RVS measurements would be beneficial for the entire semiconductor industry.

Appendix

A.1 Auxiliary Transformations and Scalings of the Weibull Distribution

Weibull CDF $F(t)$ linearization for Weibull probability plotting:

$$F(t) = 1 - \exp\left[-\left(\frac{t}{\alpha}\right)^{\beta}\right] \tag{A.1}$$

$$\Rightarrow \quad \ln\left\{-\ln\left[1 - F(t)\right]\right\} = \beta \cdot \ln\left(\frac{t}{\alpha}\right) \tag{A.2}$$

Weibull lifetime scaling for cumulative failure F:

$$t = \alpha \cdot \left[-\ln\left(1 - F\right)\right]^{\frac{1}{\beta}} \tag{A.3}$$

Conversion of Weibit Wb and cumulative failure F:

$$\mathrm{Wb} = \ln\left[-\ln\left(1 - F\right)\right] \tag{A.4}$$

$$F = 1 - \exp\left[-\exp\left(\mathrm{Wb}\right)\right] \tag{A.5}$$

Lifetime scaling of the Poisson model for Weibull distributed data:

$$t = t_{\mathrm{ref}} \cdot \left(\frac{A}{A_{\mathrm{ref}}}\right)^{-\frac{1}{\beta}} \tag{A.6}$$

Definition of the acceleration factor AF_i and common application:

$$\mathrm{AF}_{[i;\mathrm{ref}]} = \frac{\mathrm{TTF}_i}{\mathrm{TTF}_{\mathrm{ref}}} \tag{A.7}$$

$$\Rightarrow \quad \mathrm{TTF}_{\mathrm{i}} = \mathrm{AF}_{[i;\mathrm{ref}]} \cdot \mathrm{TTF}_{\mathrm{ref}} \tag{A.8}$$

A.2 Calculation Example for Multi-Dimensional Mission Profiles

This section illustrates a sample calculation of effective acceleration factors and stressors from section 4.4.2. Table A.1 shows the stresses for the standard lifetime test stress conditions and a simpler re-binned version of the Kuala Lumpur mission profile from Fig. 4.19a.

To compare the different calculation paths for Eqq. (4.26) and (4.27) and their respective effective stress results, we first intentionally and deliberately neglect all stressor interdependencies and show the inconsistent case.

Table A.1: Simple two-dimensional mission profile for OLED lighting application. Stresses have been re-binned for purpose of clarity.

Occurrence p	Temperature T	Rel. Humidity RH
Stress conditions		
100 %	85.0 °C	85.0 %
Normal operation conditions		
72 %	35.0 °C	75.0 %
4 %	35.0 °C	45.0 %
20 %	65.0 °C	15.0 %
4 %	95.0 °C	15.0 %

Hypothetically, one could separate the acceleration factors AF_T and $\mathrm{AF}_{\mathrm{RH}}$ from Eq. (4.30) in order to calculate the individual distributions of both stressors:

$$\mathrm{AF}_{T,i} = \exp\left[\frac{0.7\,\mathrm{eV}}{k_\mathrm{B}}\left(\frac{1}{T_{\mathrm{nop},i}} - \frac{1}{(85+273)\,\mathrm{K}}\right)\right] \tag{A.9}$$

$$\mathrm{AF}_{\mathrm{RH},i} = \left(\frac{\mathrm{RH}_{\mathrm{nop},i}}{85.0\,\%}\right)^{-2.7} \tag{A.10}$$

The corresponding stress conditions are 85 °C/85 % RH and the computed values are listed in Table A.2.

Table A.2: Computed acceleration factors for Peck model (see Eq. (4.30)) with presumed activation energy $Q_a = 0.7\,\text{eV}$ and Peck exponent $n = -2.7$.

p	T	AF_T	RH	AF_{RH}	$\text{AF}_T \cdot \text{AF}_{\text{RH}}$
72 %	35.0 °C	39.8	75.0 %	1.4	55.8
4 %	35.0 °C	39.8	45.0 %	5.6	221.6
20 %	65.0 °C	3.8	15.0 %	108.1	414.1
4 %	95.0 °C	0.5	15.0 %	108.1	58.4

Effective stresses are generally calculated with Eq. (4.22) and would result in these individual acceleration factors:

$$\text{AF}_{\text{eff},T} = \left(\sum_i \frac{p_i}{\text{AF}_{T,i}}\right)^{-1} = 6.9$$

$$\text{AF}_{\text{eff,RH}} = \left(\sum_i \frac{p_i}{\text{AF}_{\text{RH},i}}\right)^{-1} = 1.9$$

By rearranging Eqq. (A.9) and (A.10) with the acceleration factors of the Peck model, the effective stresses would read:

$$T_{\text{eff}} = \left(\frac{\ln\left(\text{AF}_{\text{eff},T}\right)}{\frac{0.7\,\text{eV}}{k_B}} + \frac{1}{(85+273)\,\text{K}}\right)^{-1} = 330\,\text{K} = 57.0\,°\text{C}$$

$$\text{RH}_{\text{eff}} = \left(\text{AF}_{\text{eff,RH}}\right)^{\frac{1}{-2.7}} \cdot 85.0\,\% = 66.9\,\%$$

If one would try to calculate these effective stresses for this mission profile and therefore neglect the interdependencies of temperature and relative humidity on purpose, in consequence, those effective stresses would be meaningless for the evaluation of the mission profile.

The correct consideration of the combined stressors is carried out with Eq. (4.27),

$$\text{AF}_{\text{res}} = \left(\sum_i \frac{p_i}{\text{AF}_{T,i} \cdot \text{AF}_{\text{RH},i}}\right)^{-1} = 70.1$$

$$\neq \text{AF}_{\text{eff},T} \cdot \text{AF}_{\text{eff,RH}} = 6.9 \cdot 1.9 = 13.1$$

and differs significantly from the value of the individual effective acceleration factors and Eq. (4.26).

A.3 Process Plans of TDDB Test Devices

Prozess Plan "PUF1 - 2015" J. Biba / W. Hansch

Ziel:
Bestimmung von Durchbruchsspannungen an Kapazitäten
Zuverlässigkeitsmessungen

Messung	nass-chem. Ätzen
Oxidation	Trockenätzen
Isolation	Lithographie
Reinigung	Tempern
Metallisierung	SOD

0	Wafer	Wafernummer	Orientierung	spez. Widerstand [Ωcm	Typ	Dotierung	Dicke [µm]	
		31056_3 (SXG1)	100	<0.005	n	Arsen	550	31.05.2016

1	Standard-Reinigung				
	Becherglas	SC1:	NH3 (min 25 %) : H2O2 (31 %) : H2O = 1 : 1 : 7	10 Min @ 75 °C Reg. 80 °C	31.05.2016
		SC2:	HCL (37 %) : H2O2 (31 %) : H2O = 1 : 1 : 7	10 Min @ 75 °C Reg. 80 °C	
		HF-Dip	1% HF Zeit: 30 s		31.05.2016

2	Gateoxid							
	RTP Mattson Linie 100	Rezept Temp: Typ:	8W5nm.8 750°C Nass	Vorher Dummyprozess 31056_1 Zeit: 5 min Schichtdicke: 5nm				31.05.2016
	Oxiddicke prüfen	Schichtdicke [nm]	Mitte	Flat	links	rechts	oben	
		31056_3 (SXG1)	2,7	2,7	2,6	2,7	2,7	

3	Metallisierung			
	Alu Aufdampfen	Rezept: Arbeitsdruck:	Einzelschicht-Al-WSV Controller Zieldicke: 500nm 2E-5 mbar	31.05.2016

4	Lithographie			
	Belacken	Vorbehandlung: Lack: Spincoating: Prebake:	keine AR 3740 500 U/min 5 sek / R 3 / 4000 U/min 25 sek 100 °C / 2 Min	31.05.2016
	Belichten	Maske: Hard contact Postexpuserbake:	Obere Metallisierung Exposuretime: 5.5 s 120 °C 2 min	31.05.2016
	Entwickeln	Entwickler (AR 300-475) pur Postbake (Hotplate):	Zeit: 45 sek 120 °C / 5 Min	31.05.2016
	Optische Kontrolle	31056_3 (SXG1)	ok	31.05.2016
	PNA	Temperatur: Zeit: 31056_3 als erstes geätzt	40°C 31056_3 (SXG1) 04:30 3:30 - 3:40	31.05.2016
	Lack Strippen	Isoprop/Acet	1:1 10 min	31.05.2016

5	Rückseite			
	Belacken	Lack: Spincoating: Prebake:	AR 3740 (neu) 500 U/min 5 sek / R 3 / 4000 U/min 25 sek 120 °C / 3 Min	31.05.2016
	HF-Dip	1% HF	Zeit: 1 min 40s dann hydrophob	31.05.2016
	Lack Strippen	Isoprop/Acet	1:1 10 min	31.05.2016
	Alu Aufdampfen	Rezept: Arbeitsdruck:	Einzelschicht-Al-WSV Controller Zieldicke: 500nm 2E-5 mbar	

Figure A.1: Process plan "PUF1 – 2015", created by J. Biba

Prozess Plan "MOS1 - 2017" J. Biba / A. Hirler

Ziel:
Bestimmung von Durchbruchsspannungen an Kapazitäten
Zuverlässigkeitsmessungen

Messung	nass-chem. Ätzen
Oxidation	Trockenätzen
Isolation	Lithographie
Reinigung	Tempern
Metallisierung	SOD

Nr.	Schritt	Wafernummer	Orientierung	spez. Widerstand [Ωcm]	Typ	Dotierung	Dicke [µm]
0	Wafer	01097_1 (SXF7)	100	<0.005	n	Arsen	550
		01097_2 (SXE6)	100	<0.005	n	Arsen	550

Nr.	Schritt	Prozess					
1a	Standard-Reinigung	SC1:	NH_3 (min 25 %) : H_2O_2 (31 %) : H_2O = 1 : 1 : 7			10 Min @ 75 °C Reg. 80 °C	
	Becherglas	SC2:	HCL (37 %) : H_2O_2 (31 %) : H_2O = 1 : 1 : 7			10 Min @ 75 °C Reg. 80 °C	
		HF-Dip	1% HF	Zeit:	30 s		

Nr.	Schritt	Prozess					
2a	Gateoxid	Oxofen Inotherm	Projekt: Temp: Typ:	OX850Dry.prj 850 °C 5slm O_2	Prog:	OX850Dry.prg Zeit: 10 min Schichtdicke: 4nm	
	Vor der Oxidation HF-Dip	Oxiddicke prüfen	Schichtdicke [nm]	Mitte	Flat	links	rechts	oben
			01097_1 (SXF7)					
			01097_2 (SXE6)					
3a	Metallisierung	Alu Aufdampfen	Rezept: Arbeitsdruck:	Einzelschicht-Al-WSV Controller 2E-5 mbar		Zieldicke: 500nm	

Nr.	Schritt	Prozess				
4	Lithographie	Belacken	Vorbehandlung: Lack: Spincoating: Prebake:	keine AR 3740 500 U/min 5 sek / R 3 / 4000 U/min 25 sek 100 °C / 2 Min		
		Belichten	Maske: Hard contact Postexpuserbake:	Metallisierung	Exposuretime: 5.5 s 120 °C 2 min	
		Entwickeln	Entwickler (AR 300-475) pur Postbake (Hotplate):		120 °C / 5 Min	Zeit: 45 sek
		Optische Kontrolle	01097_1 (SXF7)			
			01097_2 (SXE6)			
		PNA	Temperatur: Zeit:	40°C 3:30 - 3:40		
		Lack Strippen	Isoprop/Acet	1:1	10 min	

Nr.	Schritt	Prozess				
5	Rückseite	Belacken	Lack: Spincoating: Prebake:	AR 3740 (neu) 500 U/min 5 sek / R 3 / 4000 U/min 25 sek 120 °C / 3 Min		
		HF-Dip	1% HF	Zeit:	1 min	40s dann hydrophob
		Lack Strippen	Isoprop/Acet	1:1	10 min	
		Alu Aufdampfen	Rezept: Arbeitsdruck:	Einzelschicht-Al-WSV Controller 2E-5 mbar	Zieldicke: 500nm	

Figure A.2: Process plan "MOS1 – 2017", created by J. Biba

Prozess Plan "MOS-Guard - 2018" J. Biba / A. Hirler

Messung	nass-chem. Ätzen
Oxidation	Trockenätzen
Isolation	Lithographie
Reinigung	Tempern
Metallisierung	SOD

0	Wafer	Wafernummer	Orientierung	spez. Widerstand [Ωcm]		Typ	Dotierung	Dicke [µm]	
		06029/1	100	0,3	0,5	n	Ph	550	06.02.2019

0a	Rückseite hochdotieren								
Rückseite wird hoch n-dotiert		SOD	Aufbringen	SOD	P507	5 ml			
				Spincoating	3000 U/min für 10s				
				Bake	200°C für 10 min				
		Eindiffusion	RTP	Prog:	P507_1050°C_20s		N2:3slm	O2:1slm	
		SOD-Entfernen	NHF4 : HF : H20 = 6 : 1 : 4; vorgemischt				Soll:	01:30	
				02:00			02:00		

1	Reinigung				
		SC1:	NH3 (min 25 %) : H2O2 (31 %) : H2O = 1 : 1 : 7	10 Min @ 75 °C Reg. 80 °C	
		SC2:	HCL (37 %) : H2O2 (31 %) : H2O = 1 : 1 : 7	10 Min @ 75 °C Reg. 80 °C	

2	Oxidation							
		Oxofen Inotherm	Projekt: Temp: Typ:	OX1050Wet.prj 1050 °C Feucht, 2.8slm O2, 5 slm N2		Prog:	OX1050Wet.prg Zeit: 150 min Schichtdicke: 500 nm	
		Oxiddicke prüfen	Schichtdicke	Mitte	Flat	links	rechts	oben

3	Lithographie							
		Belacken	Vorbehandlung: Lack: Spincoating: Prebake:	HMDS AR 3740 500 U/min 5 sek / R 3 / 4000 U/min 25 sek 100 °C / 2 Min				
		Belichten	Maske: Hard contact Postexpuserbake:	Guard Ring	Exposuretime: 5.5 s 120 °C 2 min			
		Entwickeln	Entwickler (AR 300-475) pur Postbake (Hotplate):		120 °C / 5 Min		Zeit: 45 sek	
		Optische Kontrolle	Litho ok					
		BHF	NHF4 : HF : H20 = 6 : 1 : 4; vorgemischt Zeit:					
		Lack Strippen	Isoprop/Acet	1:1		10 min		
		Optische Kontrolle						

3b	Guardring hochdotieren							
		SOD	Aufbringen	SOD	B155	5 ml		
				Spincoating	3000 U/min für 10s			
				Bake	200°C für 10 min			
		Eindiffusion	RTP	Prog:	B155_1050°C_20s		N2:0slm	O2:1slm
		SOD-Entfernen	5%HF	500 ml : 50 ml; Di:HF (50%)			Soll:	01:30
							02:00	
		RIE	Rezept:	UNIBW Si3N4 CHF3-Plasma		Atzzeit:		20 s

Figure A.3: Process plan "MOS-Guard – 2018" part 1, created by J. Biba

Nr.	Schritt	Prozess	Parameter	
4	Lithographie	Belacken	Vorbehandlung: HMDS Lack: AR 3740 Spincoating: 500 U/min 5 sek / R 3 / 4000 U/min 25 sek Prebake: 100 °C / 2 Min	
		Belichten	Maske: Aktive Layer Hard contact Exposuretime: 5.5 s Postexpuserbake: 120 °C 2 min	
		Entwickeln	Entwickler (AR 300-475) pur Zeit: 45 sek Postbake (Hotplate): 120 °C / 5 Min	
		Optische Kontrolle	Litho ok	
		BHF	NHF4 : HF : H20 = 6 : 1 : 4; vorgemischt Zeit:	
		Lack Strippen	Isoprop/Acet 1:1 10 min	
		Optische Kontrolle		

Nr.	Schritt	Prozess	Parameter	
5a	Standard-Reinigung Becherglas	SC1:	NH3 (min 25 %) : H2O2 (31 %) : H2O = 1 : 1 : 7 10 Min @ 75 °C Reg. 80 °C	
		SC2:	HCL (37 %) : H2O2 (31 %) : H2O = 1 : 1 : 7 10 Min @ 75 °C Reg. 80 °C	
		HF-Dip	1% HF Zeit: 30 s	

Nr.	Schritt	Prozess	Parameter	
6a	Gateoxid Vor der Oxidation HF-Dip	RTP Mattson Linie 100	Rezept 8W5nm.8 Temp: 750°C Zeit: 10 min Typ: Nass Schichtdicke: 5nm	
		Oxiddicke prüfen	Dummy 3.7 nm	
7a	Metallisierung	Alu Aufdampfen	Rezept: Einzelschicht-Al-WSV Controller Zieldicke: 500nm Arbeitsdruck: 2E-5 mbar	

Nr.	Schritt	Prozess	Parameter	
8	Lithographie	Belacken	Vorbehandlung: keine Lack: AR 3740 Spincoating: 500 U/min 5 sek / R 3 / 4000 U/min 25 sek Prebake: 100 °C / 2 Min	
		Belichten	Maske: Metallization Hard contact Exposuretime: 5.5 s Postexpuserbake: 120 °C 2 min	
		Entwickeln	Entwickler (AR 300-475) pur Zeit: 45 sek Postbake (Hotplate): 120 °C / 5 Min	
		Optische Kontrolle		
		PNA	Temperatur: 40°C Zeit: 3:30 - 3:40	
		Lack Strippen	Isoprop/Acet 1:1 10 min	

Nr.	Schritt	Prozess	Parameter	
9	Rückseite	Belacken	Lack: AR 3740 (neu) Spincoating: 500 U/min 5 sek / R 3 / 4000 U/min 25 sek Prebake: 120 °C / 3 Min	
		HF-Dip	1% HF Zeit: 1 min 40s dann hydrophob	
		Lack Strippen	Isoprop/Acet 1:1 10 min	
		Alu Aufdampfen	Rezept: Einzelschicht-Al-WSV Controller Zieldicke: 500nm Arbeitsdruck: 2E-5 mbar	

Figure A.4: Process plan "MOS-Guard – 2018" part 2, created by J. Biba

List of Abbreviations

AEC	Automotive Electronics Council	30
AD	Anderson-Darling	25
AF	acceleration factor	5
AFR	average failure rate	5
ALT	accelerated lifetime testing	30
BEOL	back end of line	38
CDF	cumulative distribution function	4
CE	cumulative exposure	62
CST	cyclical stress testing	33
CVS	constant voltage stress	43
DUT	device under test	30
EDF	empirical distribution function	4
EFR	early failure rate	6
FD-SOI	fully depleted silicon on insulator	47
FEOL	front end of line	38
FR	failure rate	4
fWLR	fast wafer level reliability	30
GOF	goodness-of-fit	11
HALT	highly accelerated lifetime testing	30
HTOL	high temperature operating life	30
IFR	intrinsic failure rate	7
JEDEC	Joint Electron Device Engineering Council	31
FIT	failures in time	1
MLE	maximum likelihood estimation	23
MOL	middle of line	38
MOS	metal–oxide–semiconductor	40
MOSFET	metal–oxide–semiconductor field-effect transistor	38
MTBF	mean time between failures	5

Bibliography

[Aal06] Andreas Aal. "Fast Prediction Of Gate Oxide Reliability - Application Of The Cumulative Damage Principle For Transforming V-Ramp Breakdown Distributions Into TDDB Failure Distributions". In: *2006 IEEE International Integrated Reliability Workshop Final Report*. IEEE, Oct. 2006, pp. 182–185. ISBN: 1-4244-0296-4. DOI: 10.1109/IRWS.2006.305241. URL: http://ieeexplore.ieee.org/document/4098718/ (cit. on p. 105).

[Aal07] Andreas Aal. "TDDB Data Generation for Fast Lifetime Projections Based on V-Ramp Stress Data". In: *IEEE Transactions on Device and Materials Reliability* 7.2 (June 2007), pp. 278–284. ISSN: 1530-4388. DOI: 10.1109/TDMR.2007.901091. URL: http://ieeexplore.ieee.org/document/4295074/ (cit. on pp. 64, 105).

[Aal08] Andreas Aal. "A Comparison Between V-Ramp TDDB Techniques For Reliability Evaluation". In: *2008 IEEE International Integrated Reliability Workshop Final Report*. IEEE, Oct. 2008, pp. 133–136. ISBN: 978-1-4244-2194-7. DOI: 10.1109/IRWS.2008.4796104. URL: http://ieeexplore.ieee.org/document/4796104/ (cit. on p. 105).

[Agi13] Agilent Technologies. *Agilent B1500A Semiconductor Device Analyzer: User's Guide*. Agilent Technologies, 2013, p. 202 (cit. on p. 42).

[Agu19] Fernando Leonel Aguirre et al. "Spatio-Temporal Defect Generation Process in Irradiated HfO2 MOS Stacks: Correlated Versus Uncorrelated Mechanisms". In: *2019 IEEE International Reliability Physics Symposium (IRPS)*. IEEE, Mar. 2019, pp. 1–8. ISBN: 978-1-5386-9504-3. DOI: 10.1109/IRPS.2019.8720539. URL: https://ieeexplore.ieee.org/document/8720539/ (cit. on p. 39).

[Ala02a] M. A. Alam, B. E. Weir, and P. J. Silverman. "A study of soft and hard breakdown - Part I: Analysis of statistical percolation conductance". In: *IEEE Transactions on Electron Devices* 49.2 (2002), pp. 232–238. ISSN: 00189383. DOI: 10.1109/16.981212. URL: https://ieeexplore.ieee.org/document/981212 (cit. on pp. 38, 39).

[Ala02b] M. A. Alam, B. E. Weir, and P. J. Silverman. "A study of soft and hard breakdown - Part II: Principles of area, thickness, and voltage scaling". In: *IEEE Transactions on Electron Devices* 49.2 (2002), pp. 239–246. ISSN: 00189383. DOI: 10.1109/16.981213. URL: http://ieeexplore.ieee.org/document/981213/ (cit. on p. 38).

[And52] T. W. Anderson and D. A. Darling. "Asymptotic Theory of Certain "Goodness of Fit" Criteria Based on Stochastic Processes". In: *The Annals of Mathematical Statistics* 23.2 (June 1952), pp. 193–212. ISSN: 0003-4851. DOI: 10.1214/aoms/1177729437. URL: http://projecteuclid.org/euclid.aoms/1177729437 (cit. on p. 25).

[And54] T. W. Anderson and D. A. Darling. "A Test of Goodness of Fit". In: *Journal of the American Statistical Association* 49.268 (Dec. 1954), p. 765. ISSN: 01621459. DOI: 10.2307/2281537. URL: https://www.jstor.org/stable/2281537 (cit. on p. 25).

[Aut14] Automotive Electronics Council. "AEC-Q100 - Failure Mechanism Based Stress Test Qualification for Integrated Circuits". In: *AEC - Q100 Rev-H.* 2014. URL: http://www.aecouncil.com/Documents/AEC_Q100_Rev_H_Base_Document.pdf (cit. on pp. 30, 60, 61).

[Bai18] Brian Bailey. *New Market Drivers - Part 2.* 2018. URL: https://semiengineering.com/new-market-drivers-2/ (visited on 07. 05. 2018) (cit. on p. 1).

[Bha89] G. K. Bhattacharyya and Zanzawi Soejoeti. "A tampered failure rate model for step-stress accelerated life test". In: *Communications in Statistics - Theory and Methods* 18.5 (Jan. 1989), pp. 1627–1643. ISSN: 0361-0926. DOI: 10.1080/03610928908829990. URL: http://www.tandfonline.com/doi/abs/10.1080/03610928908829990 (cit. on pp. 62, 63, 66, 67, 74).

[Bro12] Ilja N. Bronštejn. *Taschenbuch der Mathematik.* 8th ed. Frankfurt am Main: Harri Deutsch, 2012. ISBN: 978-3-8171-2008-6 (cit. on pp. 10, 11, 111).

[Cox72] David Roxbee Cox. "Regression Models and Life-Tables". In: *Journal of the Royal Statistical Society. Series B (Methodological)* 34.2 (1972), pp. 187–220. ISSN: 00359246. URL: http://www.jstor.org/stable/2985181 (cit. on p. 67).

[Cro15] K. Croes et al. "Impact of process variability on BEOL TDDB lifetime model assessment". In: *2015 IEEE International Reliability Physics Symposium.* 32. IEEE, Apr. 2015, BD.5.1–BD.5.5. ISBN: 978-1-4673-7362-3. DOI: 10.1109/IRPS.2015.7112777. URL: https://ieeexplore.ieee.org/document/7112777 (cit. on p. 50).

[DAg86] Ralph B. D'Agostino and Michael A. Stephens. *Goodness-of-fit techniques.* 1. print. Statistics; vol. 68. New York u.a.: M. Dekker, 1986, p. 560. ISBN: 0824774876 (cit. on pp. 25, 27).

[Deg00] Robin Degraeve, Ben Kaczer, and Guido Groeseneken. "Reliability: a possible showstopper for oxide thickness scaling?" In: *Semiconductor Science and Technology* 15.5 (May 2000), pp. 436–444. ISSN: 0268-1242. DOI: 10.1088/0268-1242/15/5/302. URL: https://iopscience.iop.org/article/10.1088/0268-1242/15/5/302 (cit. on pp. 39, 40).

[DeG79] Morris H. DeGroot and Prem K. Goel. "Bayesian estimation and optimal designs in partially accelerated life testing". In: *Naval Research Logistics Quarterly* 26.2 (June 1979), pp. 223–235. ISSN: 00281441. DOI: 10.1002/nav.3800260204. URL: https://onlinelibrary.wiley.com/doi/abs/10.1002/nav.3800260204 (cit. on pp. 62, 63, 65).

[Eft07] E. Efthymiou et al. "Reliability nano-characterization of thin SiO2 and HfSixOy/SiO2 gate stacks". In: *Microelectronic Engineering* 84.9-10 (Sept. 2007), pp. 2290–2293. ISSN: 01679317. DOI: 10.1016/j.mee.2007.04.060. URL: https://www.sciencedirect.com/science/article/abs/pii/S0167931707004108 (cit. on p. 39).

[Ham15] M. S. Hamada. "Bayesian Analysis of Step-Stress Accelerated Life Tests and Its Use in Planning". In: *Quality Engineering* 27.3 (July 2015), pp. 276–282. ISSN: 0898-2112. DOI: 10.1080/08982112.2015.1038357. URL: https://www.tandfonline.com/doi/full/10.1080/08982112.2015.1038357 (cit. on p. 62).

[Har84] H. Leon Harter. "Another look at plotting positions". In: *Communications in Statistics - Theory and Methods* 13.13 (Jan. 1984), pp. 1613–1633. ISSN: 0361-0926. DOI: 10.1080/03610928408828781. URL: http://www.tandfonline.com/doi/abs/10.1080/03610928408828781 (cit. on pp. 16, 17).

[Hu99] Chenming Hu and Qiang Lu. "A unified gate oxide reliability model". In: *1999 IEEE International Reliability Physics Symposium Proceedings. 37th Annual (Cat. No.99CH36296)*. IEEE, 1999, pp. 47–51. ISBN: 0-7803-5220-3. DOI: 10.1109/RELPHY.1999.761591. URL: https://ieeexplore.ieee.org/document/761591 (cit. on pp. 31, 50).

[Jac97] J. C. Jackson et al. "Nonuniqueness of time-dependent-dielectric-breakdown distributions". In: *Applied Physics Letters* 71.25 (1997), p. 3682. ISSN: 00036951. DOI: 10.1063/1.120480. URL: https://aip.scitation.org/doi/abs/10.1063/1.120480 (cit. on pp. 40, 43).

[JED01] JEDEC. "Procedure for the Wafer-Level Testing of Thin Dielectrics". In: *JEDEC Standard JESD35-A*. Apr. 2001. URL: https://www.jedec.org/sites/default/files/docs/jesd35a.pdf (cit. on pp. 43, 44, 103).

[JED03] JEDEC. "Procedure for Characterizing Time-Dependent Dielectric Breakdown of Ultra-Thin Gate Dielectrics". In: *JEDEC Standard JESD92*. Aug. 2003. URL: https://www.jedec.org/system/files/docs/JESD92.pdf (cit. on pp. 31, 39, 43, 44, 103, 105).

[JED10] JEDEC. "JEDEC Procedure for the Evaluation of Low-k/Metal Inter/Intra-Level Dielectric Integrity". In: *JEDEC Standard JEP159*. Aug. 2010. URL: https://www.jedec.org/sites/default/files/docs/JEP159.pdf (cit. on p. 48).

[JED16] JEDEC. "Failure Mechanisms and Models for Semiconductor Devices". In: *JEDEC Publication JEP122H*. Sept. 2016. URL: https://www.jedec.org/system/files/docs/JEP122H.pdf (cit. on pp. 9, 37, 47, 50, 51, 76, 83).

[Kan14] W. Kanert. "Robustness Validation – A physics of failure based approach to qualification". In: *Microelectronics Reliability* 54.9-10 (Sept. 2014), pp. 1648–1654. ISSN: 00262714. DOI: 10.1016/j.microrel.2014.07.010. URL: https://www.sciencedirect.com/science/article/abs/pii/S0026271414002066 (cit. on p. 61).

[Kap58] E. L. Kaplan and Paul Meier. "Nonparametric Estimation from Incomplete Observations". In: *Journal of the American Statistical Association* 53.282 (June 1958), p. 457. ISSN: 01621459. DOI: 10.2307/2281868. URL: http://www.jstor.org/stable/2281868 (cit. on p. 20).

[Ker06a] A. Kerber et al. "From wafer-level gate-oxide reliability towards ESD failures in advanced CMOS technologies". In: *IEEE Transactions on Electron Devices* 53.4 (Apr. 2006), pp. 917–920. ISSN: 0018-9383. DOI: 10.1109/TED.2006.870517. URL: http://ieeexplore.ieee.org/document/1610929/ (cit. on p. 113).

[Ker06b] A. Kerber et al. "Impact of failure criteria on the reliability prediction of CMOS devices with ultrathin gate oxides based on voltage ramp stress". In: *IEEE Electron Device Letters* 27.7 (July 2006), pp. 609–611. ISSN: 0741-3106. DOI: 10.1109/LED.2006.877710. URL: http://ieeexplore.ieee.org/document/1644842/ (cit. on p. 117).

[Ker07] A. Kerber et al. "Reliability screening of high-k dielectrics based on voltage ramp stress". In: *Microelectronics Reliability* 47.4-5 (Apr. 2007), pp. 513–517. ISSN: 00262714. DOI: 10.1016/j.microrel.2007.01.030. URL: https://www.sciencedirect.com/science/article/abs/pii/S0026271407000364 (cit. on pp. 106, 113, 117, 121).

[Key16] Keysight Technologies. *Keysight EasyEXPERT Software User's Guide Volume 1*. Vol. B1540-9000. Keysight Technologies, 2016. URL: http://literature.cdn.keysight.com/litweb/pdf/B1540-90000.pdf?id=819567 (cit. on p. 102).

[Key17] Keysight Technologies. *The Parametric Measurement Handbook*. 4th ed. Keysight Technologies, Dec. 2017. ISBN: 5992-2508EN. URL: https://literature.cdn.keysight.com/litweb/pdf/5992-2508EN.pdf?id=2941515 (cit. on pp. 102, 105).

[Kwe99] Kwee-Poo Yeo and Loon-Ching Tang. "Planning step-stress life-test with a target acceleration-factor". In: *IEEE Transactions on Reliability* 48.1 (Mar. 1999), pp. 61–67. ISSN: 00189529. DOI: 10.1109/24.765928. URL: http://ieeexplore.ieee.org/document/765928/ (cit. on p. 62).

[Li07] Chenhua Li and Nasser Fard. "Optimum Bivariate Step-Stress Accelerated Life Test for Censored Data". In: *IEEE Transactions on Reliability* 56.1 (Mar. 2007), pp. 77–84. ISSN: 0018-9529. DOI: 10.1109/TR.2006.890897. URL: http://ieeexplore.ieee.org/document/4118424/ (cit. on p. 62).

[Lin01] B. P. Linder et al. "Transistor-limited constant voltage stress of gate dielectrics". In: *2001 Symposium on VLSI Technology. Digest of Technical Papers (IEEE Cat. No.01 CH37184)*. C. Japan Soc. Appl. Phys, 2001, pp. 93–94. ISBN: 4-89114-012-7. DOI: 10.1109/VLSIT.2001.934965. URL: http://ieeexplore.ieee.org/document/934965/ (cit. on p. 40).

[Mad93] Mohamed T. Madi. "Multiple step-stress accelerated life test: the tampered failure rate model". In: *Communications in Statistics - Theory and Methods* 22.9 (Jan. 1993), pp. 295–306. ISSN: 0361-0926. DOI: 10.1080/03610928308831174. URL: http://www.tandfonline.com/doi/abs/10.1080/03610928308831174 (cit. on p. 69).

[Mat18] MathWorks. *MATLAB*. Natick, Massachusetts, 2018. URL: https://de.mathworks.com/help/stats/probplot.html (visited on 20.04.2018) (cit. on p. 17).

[Mat20] MathWorks. *MATLAB*. Natick, Massachusetts, 2020. URL: https://de.mathworks.com/help/symbolic/lambertw.html (visited on 11.08.2020) (cit. on p. 109).

[McP00] J. W. McPherson, R. B. Khamankar, and A. Shanware. "Complementary model for intrinsic time-dependent dielectric breakdown in SiO2 dielectrics". In: *Journal of Applied Physics* 88.9 (2000), p. 5351. ISSN: 00218979. DOI: 10.1063/1.1318369. URL: https://aip.scitation.org/doi/abs/10.1063/1.1318369 (cit. on pp. 31, 50).

[McP12] J. W. McPherson. "Time dependent dielectric breakdown physics – Models revisited". In: *Microelectronics Reliability* 52.9-10 (Sept. 2012), pp. 1753–1760. ISSN: 00262714. DOI: 10.1016/j.microrel.2012.06.007. URL: http://linkinghub.elsevier.com/retrieve/pii/S0026271412001916 (cit. on pp. 38, 47, 50).

[McP19] J. W. McPherson. *Reliability Physics and Engineering*. 3rd ed. Cham: Springer International Publishing, July 2019. ISBN: 978-3-319-93682-6. DOI: 10.1007/978-3-319-93683-3. URL: http://link.springer.com/10.1007/978-3-319-93683-3 (cit. on pp. 3, 7, 9–11, 13, 17, 38, 39, 47, 50, 51, 62, 107, 113).

[McP85] J. W. McPherson and D. A. Baglee. "Acceleration Factors for Thin Gate Oxide Stressing". In: *23rd International Reliability Physics Symposium.* Vol. 132. 8. IEEE, Mar. 1985, pp. 1–5. ISBN: 0735-0791. DOI: 10.1109/IRPS.1985.362066. URL: https://ieeexplore.ieee.org/document/4208593/ (cit. on p. 117).

[Mil83] Robert Miller and Wayne Nelson. "Optimum Simple Step-Stress Plans for Accelerated Life Testing". In: *IEEE Transactions on Reliability* R-32.1 (Apr. 1983), pp. 59–65. ISSN: 0018-9529. DOI: 10.1109/TR.1983.5221475. URL: http://ieeexplore.ieee.org/document/5221475/ (cit. on p. 62).

[Min19] Minitab. *Methods and formulas for probability plot in Distribution ID Plot (Arbitrary Censoring).* 2019. URL: https://support.minitab.com/en-us/minitab/18/help-and-how-to/modeling-statistics/reliability/how-to/distribution-id-plot-arbitrary-censoring/methods-and-formulas/probability-plot/ (visited on 25.10.2019) (cit. on p. 17).

[Nab93] Seiji Nabeya. "Coincidence of two failure rate models". In: *Communications in Statistics - Theory and Methods* 22.3 (Jan. 1993), pp. 781–785. ISSN: 0361-0926. DOI: 10.1080/03610929308831055. URL: http://www.tandfonline.com/doi/abs/10.1080/03610929308831055 (cit. on p. 74).

[Nel80] Wayne B. Nelson. "Accelerated Life Testing – Step-Stress Models and Data Analyses". In: *IEEE Transactions on Reliability* R-29.2 (June 1980), pp. 103–108. ISSN: 0018-9529. DOI: 10.1109/TR.1980.5220742. URL: http://ieeexplore.ieee.org/document/5220742/ (cit. on pp. 62–64).

[Nel82] Wayne B. Nelson. *Applied Life Data Analysis.* Vol. 577. John Wiley & Sons, 1982, p. 650. ISBN: 9780471725220 (cit. on pp. 8–11).

[Nel90] Wayne B. Nelson. *Accelerated Testing.* Wiley Series in Probability and Statistics. Hoboken, NJ, USA: John Wiley & Sons, Inc., Feb. 1990. ISBN: 9780470316795. DOI: 10.1002/9780470316795. URL: http://doi.wiley.com/10.1002/9780470316795 (cit. on pp. 17, 19, 33, 62).

[NIS13] NIST and SEMATECH. *e-Handbook of Statistical Methods.* 2013. URL: http://www.itl.nist.gov/div898/handbook/ (visited on 11.01.2017) (cit. on pp. 3, 17, 22–25).

[Oka17] Kenji Okada, Masayuki Kamei, and Shigeyuki Ohno. "Reconsideration of Dielectric Breakdown Mechanism of Gate Dielectrics on Basis of Dominant Carrier Change Model". In: *IEEE Transactions on Electron Devices* 64.11 (Nov. 2017), pp. 4386–4392. ISSN: 0018-9383. DOI: 10.1109/TED.2017.2747580. URL: http://ieeexplore.ieee.org/document/8036396/ (cit. on p. 50).

[Ori17] Origin. *Wahrscheinlichkeitsdiagramm und Q-Q-Diagramm.* 2017. URL: http://www.originlab.com/doc/Origin-Help/ProbPlot-QQPlot (visited on 24. 01. 2017) (cit. on p. 17).

[Pec86] D. Stewart Peck. "Comprehensive Model for Humidity Testing Correlation". In: *24th International Reliability Physics Symposium.* IEEE, Apr. 1986, pp. 44–50. DOI: 10.1109/IRPS.1986.362110. URL: http://ieeexplore.ieee.org/document/4208640/ (cit. on p. 95).

[Pen10] Chien-Yu Peng and Sheng-Tsaing Tseng. "Progressive-Stress Accelerated Degradation Test for Highly-Reliable Products". In: *IEEE Transactions on Reliability* 59.1 (Mar. 2010), pp. 30–37. ISSN: 0018-9529. DOI: 10.1109/TR.2010.2040769. URL: http://ieeexplore.ieee.org/document/5418918/ (cit. on p. 62).

[Por03] M. Porti et al. "Atomic force microscope topographical artifacts after the dielectric breakdown of ultrathin SiO2 films". In: *Surface Science* 532-535 (June 2003), pp. 727–731. ISSN: 00396028. DOI: 10.1016/S0039-6028(03)00150-X. URL: https://www.sciencedirect.com/science/article/abs/pii/S003960280300150X (cit. on p. 39).

[Rah11] Nilufa Rahim, Ernest Y. Wu, and Durgamadhab Misra. "Investigation of progressive breakdown and non-Weibull failure distribution of high-k and SiO2 dielectric by ramp voltage stress". In: *2011 International Reliability Physics Symposium.* IEEE, Apr. 2011, GD.2.1–GD.2.6. ISBN: 978-1-4244-9113-1. DOI: 10.1109/IRPS.2011.5784579. URL: http://ieeexplore.ieee.org/document/5784579/ (cit. on p. 113).

[Rel15] ReliaSoft Corporation. *Life Data Analysis Reference.* ReliaSoft Corporation, 2015, pp. 103–154. URL: http://www.synthesisplatform.net/references/Life_Data_Analysis_Reference.pdf (cit. on pp. 23, 24).

[Ros96] Elyse Rosenbaum, Josef C. King, and Chenming Hu. "Accelerated testing of SiO2 reliability". In: *IEEE Transactions on Electron Devices* 43.1 (1996), pp. 70–80. ISSN: 00189383. DOI: 10.1109/16.477595. URL: http://ieeexplore.ieee.org/document/477595/ (cit. on p. 117).

[Sha14] Naijun Sha and Rong Pan. "Bayesian analysis for step-stress accelerated life testing using weibull proportional hazard model". In: *Statistical Papers* 55.3 (Aug. 2014), pp. 715–726. ISSN: 0932-5026. DOI: 10.1007/s00362-013-0521-2. URL: http://link.springer.com/10.1007/s00362-013-0521-2 (cit. on pp. 67, 74, 86).

[Sha83] Moshe Shaked and Nozer D. Singpurwalla. "Inference for step-stress accelerated life tests". In: *Journal of Statistical Planning and Inference* 7.4 (June 1983), pp. 295–306. ISSN: 03783758. DOI: 10.1016/0378-3758(83)90001-0. URL: https://www.sciencedirect.com/science/article/abs/pii/0378375883900010 (cit. on p. 66).

[Som02] Sergio S. Sombra et al. "A percolation based dielectric breakdown model with randomic changes in the dielectric constant". In: *Physica A: Statistical Mechanics and its Applications* 305.3-4 (Mar. 2002), pp. 351–359. ISSN: 03784371. DOI: 10.1016/S0378-4371(01)00615-X. URL: https://www.sciencedirect.com/science/article/abs/pii/S037843710100615X (cit. on pp. 38, 39, 83, 88).

[Ste74] M. A. Stephens. "EDF Statistics for Goodness of Fit and Some Comparisons". In: *Journal of the American Statistical Association* 69.347 (Sept. 1974), p. 730. ISSN: 01621459. DOI: 10.2307/2286009. URL: https://www.jstor.org/stable/2286009 (cit. on p. 25).

[Str09] Alvin W. Strong et al. *Reliability Wearout Mechanisms in Advanced CMOS Technologies*. Hoboken, NJ, USA: John Wiley & Sons, Inc., Aug. 2009. ISBN: 9780470455265. DOI: 10.1002/9780470455265. URL: http://doi.wiley.com/10.1002/9780470455265 (cit. on pp. 3–5, 7–10, 13, 16, 19, 20, 24, 31, 38, 39, 47, 50, 51, 121).

[Sue93] J. S. Suehle. "Reproducibility of JEDEC Standard Current and Voltage Ramp Test Procedures for Thin-Dielectric Breakdown Characterization". In: *1993 IEEE International Integrated Reliability Workshop Final Report*. IEEE, 1993, pp. 22–34. DOI: 10.1109/IRWS.1993.666288. URL: http://ieeexplore.ieee.org/document/666288/ (cit. on p. 103).

[Sze06] S. M. Sze and Kwok K. Ng. *Physics of Semiconductor Devices*. Hoboken, NJ, USA: John Wiley & Sons, Inc., Oct. 2006. ISBN: 9780470068328. DOI: 10.1002/0470068329. URL: http://doi.wiley.com/10.1002/0470068329 (cit. on p. 45).

[Tan96] L. C. Tang et al. "Analysis of step-stress accelerated-life-test data: a new approach". In: *IEEE Transactions on Reliability* 45.1 (Mar. 1996), pp. 69–74. ISSN: 00189529. DOI: 10.1109/24.488919. URL: http://ieeexplore.ieee.org/document/488919/ (cit. on p. 62).

[Wan04] Ronghua Wang and Heliang Fei. "Conditions for the coincidence of the TFR, TRV and CE models". In: *Statistical Papers* 45.3 (July 2004), pp. 393–412. ISSN: 0932-5026. DOI: 10.1007/BF02777579. URL: http://link.springer.com/10.1007/BF02777579 (cit. on p. 74).

[Wei51] Waloddi Weibull. "A Statistical Distribution Function of Wide Applicability". In: *ASME Journal of Applied Mechanics* 18.3 (1951), pp. 293–297 (cit. on pp. 8, 9).

[Wol20] Wolfram Research. *Mathematica*. Champaign, Illinois, 2020. URL: https://reference.wolfram.com/language/ref/ProductLog.html (visited on 11.08.2020) (cit. on p. 109).

[Wol85] D. R. Wolters and J. J. Van der Schoot. "Dielectric Breakdown in MOS Devices. Part I: Defect-Related and Intrinsic Breakdown". In: *Philips Journal of Research* 40.3 (1985), pp. 115–136. ISSN: 01655817 (cit. on p. 117).

[Wu02] Ernest Y. Wu, Jordi Suñé, and W. Lai. "On the weibull shape factor of intrinsic breakdown of dielectric films and its accurate experimental determination-part II: experimental results and the effects of stress conditions". In: *IEEE Transactions on Electron Devices* 49.12 (Dec. 2002), pp. 2141–2150. ISSN: 0018-9383. DOI: 10.1109/TED.2002.805603. URL: http://ieeexplore.ieee.org/document/1177978/ (cit. on p. 39).

[Wu09a] Ernest Y. Wu and Jordi Suñé. "On Voltage Acceleration Models of Time to Breakdown — Part I: Experimental and Analysis Methodologies". In: *IEEE Transactions on Electron Devices* 56.7 (2009), pp. 1433–1441. ISSN: 0018-9383. DOI: 10.1109/TED.2009.2021721. URL: http://ieeexplore.ieee.org/document/5072256/ (cit. on pp. 50, 117).

[Wu09b] Ernest Y. Wu and Jordi Suñé. "On Voltage Acceleration Models of Time to Breakdown — Part II: Experimental Results and Voltage Dependence of Weibull Slope in the FN Regime". In: *IEEE Transactions on Electron Devices* 56.7 (2009), pp. 1442–1450. ISSN: 0018-9383. DOI: 10.1109/TED.2009.2021725. URL: http://ieeexplore.ieee.org/document/5071301/ (cit. on p. 50).

[Wu09c] Ernest Y. Wu and Jordi Suñé. "Towards a viable TDDB reliability assessment methodology: From breakdown physics to circuit failure". In: *2009 16th IEEE International Symposium on the Physical and Failure Analysis of Integrated Circuits*. IEEE, July 2009, pp. 63–70. ISBN: 978-1-4244-3911-9. DOI: 10.1109/IPFA.2009.5232698. URL: https://ieeexplore.ieee.org/document/5232698 (cit. on p. 40).

[Wu20] C. Wu et al. "Conduction and Breakdown Mechanisms in Low-k Spacer and Nitride Spacer Dielectric Stacks in Middle of Line Interconnects". In: *2020 IEEE International Reliability Physics Symposium (IRPS)*. Vol. 2020-April. IEEE, Apr. 2020, pp. 1–6. ISBN: 978-1-7281-3199-3. DOI: 10.1109/IRPS45951.2020.9128328. URL: https://ieeexplore.ieee.org/document/9128328/ (cit. on p. 37).

[Xio99] Chengjie Xiong and G. A. Milliken. "Step-stress life-testing with random stress-change times for exponential data". In: *IEEE Transactions on Reliability* 48.2 (June 1999), pp. 141–148. ISSN: 00189529. DOI: 10.1109/24.784272. URL: http://ieeexplore.ieee.org/document/784272/ (cit. on p. 62).

[Zha05] Wenbiao Zhao and Elsayed A. Elsayed. "A general accelerated life model for step-stress testing". In: *IIE Transactions* 37.11 (Nov. 2005), pp. 1059–1069. ISSN: 0740-817X. DOI: 10.1080/07408170500232396. URL: http://www.tandfonline.com/doi/abs/10.1080/07408170500232396 (cit. on pp. 62, 64).

[ZVE15] ZVEI. *Handbook for Robustness Validation of Semiconductor Devices in Automotive Applications.* May 2015. URL: https://www.zvei.org/fileadmin/user_upload/Presse_und_Medien/Publikationen/2015/mai/Handbook_for_Robustness_Validation_of_Semiconductor_Devices_in_Automotive_Applications__3rd_edition_/Robustness-Validation-Semiconductor-2015.pdf (cit. on p. 27).

Publications

(1) A. Hirler, J. Biba, A. Alsioufy, T. Lehndorff, T. Sulima, H. Lochner, U. Abelein, and W. Hansch. "Umsetzung von Mission-Profiles in effektive Stressniveaus und -zeiten für die Zuverlässigkeit von Halbleiterbauelementen". In: *edaWorkshop17*, 08.05–10.05.2017, Dresden.

(2) T. Lehndorff, A. Alsioufy, A. Hirler, T. Sulima, H. Lochner, and W. Hansch. "Strukturierte Halbleitertechnologiebewertung als Innovationsbeschleuniger im Automobil". In: *edaWorkshop17*, 08.05–10.05.2017, Dresden.

(3) A. Alsioufy, A. Hirler, T. Lehndorff, T. Sulima, H. Lochner, and W. Hansch. "Entwicklung einer Technologie-BlackBox für die strukturierte Technologiebewertung in Automotive Anwendungen". In: *edaWorkshop17*, 08.05–10.05.2017, Dresden.

(4) T. Lehndorff, A. Alsioufy, A. Hirler, T. Sulima, H. Lochner, and W. Hansch. "autoSWIFT: Absicherung neuer Automotive-Anforderungen durch strukturierte Halbleiter-Technologiebewertung". In: *AmE 2017 – Automotive meets Electronics, 8th GMM-Symposium*, 07.–08.03.2017, Dortmund. VDE, GMM-Fb. 87: AmE 2017, ISBN: 978-3-8007-4369-8. URL: https://ieeexplore.ieee.org/document/8335219

(5) T. Lehndorff, M. Eberhardt, H. Lochner, G. Jerke, A. Alsioufy, A. Hirler, T. Sulima, and W. Hansch. "REConf 2017: Wie muss sich das Zusammenspiel von Requirements Engineering und den anderen Systems Engineering-Disziplinen im Produktentstehungsprozess entlang der Wertschöpfungskette der Automobilindustrie für die Zukunft aufstellen?". In: *REConf* (2017), München. URL: https://www.hood-group.com/reconf/archiv/reconf-2017/agenda

(6) A. Hirler, J. Biba, A. Alsioufy, T. Lehndorff, T. Sulima, H. Lochner, U. Abelein, and W. Hansch. “Evaluation of effective stress times and stress levels from mission profiles for semiconductor reliability”. *28th European Symposium on Reliability of Electronic Devices, Failure Physics and Analysis ESREF 2017*, 25.09.–28.09.2017, Bordeaux, France. In: *Microelectronics Reliability* 76–77 (2017), pp. 38–41. DOI: 10.1016/j.microrel.2017.06.022

(7) A. Hirler, A. Alsioufy, T. Lehndorff, H. Lochner, S. Simon, T. Sulima, and W. Hansch. “Mehrdimensionale Mission-Profiles – Berechnung effektiver Stressoren”. In: *edaWorkshop18*, 16.–17.05.2018, Hannover.

(8) A. Alsioufy, A. Hirler, T. Lehndorff, T. Sulima, H. Lochner, S. Simon, and W. Hansch. “Die Technologie-BlackBox: Pionierwerkzeug für die strukturierte Halbleitertechnologiebewertung in Automotive Anwendungen”. In: *edaWorkshop18*, 16.–17.05.2018, Hannover.

(9) T. Lehndorff, A. Alsioufy, A. Hirler, T. Sulima, H. Lochner, S. Simon, and W. Hansch. “Die Technologie-BlackBox: Anwendungsszenarien und Use-Cases”. In: *edaWorkshop18*, 16.–17.05.2018, Hannover.

(10) A. Hirler, A. Alsioufy, J. Biba, T. Lehndorff, D. Lipp, H. Lochner, M. Siddabathula, S. Simon, T. Sulima, M. Wiatr, and W. Hansch. “Alternating Temperature Stress and Deduction of Effective Stress Levels from Mission Profiles for Semiconductor Reliability”. In: *IEEE International Reliability Physics Symposium IRPS* (2019), 31.03.–04.04.2019, Monterey, California, USA. DOI: 10.1109/IRPS.2019.8720536

(11) A. Hirler, A. Alsioufy, J. Biba, T. Lehndorff, H. Lochner, S. Simon, T. Sulima, W. Thomas, and W. Hansch. “Effective and combined stressors from multi-dimensional mission profiles for semiconductor reliability”. *30th European Symposium on Reliability of Electronic Devices, Failure Physics and Analysis ESREF 2019*, 23.09.–26.09.2019, Toulouse, France. In: *Microelectronics Reliability* 100–101C (2019), p. 113323. DOI: 10.1016/j.microrel.2019.06.015

(12) A. Hirler, “How to deal with Mission-Profiles as Reliability Requirements?”. In: *VDE ITG MN 5.6 – Fachtagung (fWLR / Wafer Level Reliability, Zuverlässigkeits-Simulation & Qualifikation)*, 27.05.–29.05.2019, Dresden.

(13) A. Hirler, “Mission Profiles as a Key Enabler for Semiconductor Reliability Qualification along the Automotive Supply Chain”. In: *Infineon University Evening*, 24.10.2019, Neubiberg.

(14) A. Alsioufy, A. Hirler, T. Lehndorff, T. Sulima, H. Lochner, S. Simon, M. Siddabathula, M. Wiatr, and W. Hansch. "Technology Black Box: A Pioneering Tool for Semiconductor Technology Development in the Automotive Industry". In: *AtheneForschung*, Universität der Bundeswehr München (2020). DOI: 10.18726/2020_1

(15) T. Lehndorff, U. Abelein, A. Alsioufy, A. Hirler, T. Sulima, S. Simon, H. Lochner, and W. Hansch. "Extended lifetime qualification concepts for automotive semiconductor components". In: *AtheneForschung*, Universität der Bundeswehr München (2020). DOI: 10.18726/2020_2

(16) A. Hirler, J. Biba, D. Lipp, H. Lochner, M. Siddabathula, S. Simon, T. Sulima, M. Wiatr, and W. Hansch. "Experimental reliability study of cumulative damage models on state-of-the-art semiconductor technologies for step-stress tests and mission profile stresses". In: *J. Vac. Sci. Technol. B* 38 (2020), p. 064001. DOI: 10.1116/6.0000504

Danksagung

Abschließend möchte ich mich bei all denjenigen bedanken, die direkt oder indirekt zum Gelingen dieser Arbeit beigetragen und mich die letzten Jahre über begleitet haben. Im Besonderen geht mein herzlicher Dank an:

Univ.-Prof. Dr.-Ing. Walter Hansch, meinen Doktorvater, für sein Vertrauen und seinen Einsatz. Er bot mir diese interessante Themenstellung an und war immer darauf bedacht mir die wissenschaftliche Freiheit zu ermöglichen, meine Forschung nach Interesse und Neugier selbst ausgestalten zu können. Seine persönliche und individuelle Betreuung und Unterstützung haben diese Doktorarbeit sehr bereichert. Durch seinen warmherzigen Umgang und die herzliche Aufnahme in die Instituts-Familie war die Arbeitsatmosphäre durchwegs konstruktiv, harmonisch und von Hilfsbereitschaft geprägt.

Dr.-Ing. Josef Biba für die hervorragende technologische Unterstützung während der gesamten Arbeit, angefangen beim Rocket-Run für die ersten Testwafer, über die Prozessplanung, das Maskenzeichnen und die Prozessierung der weiteren Proben und der Einführung in das Messlabor des Institutes, bis hin zu den vielen fruchtbaren Diskussionen zum großen Themenkomplex der Halbleiterei. Ebenfalls einen besonderen Dank für die vielen Stunden des Korrekturlesens dieser Arbeit.

Dr.-Ing. Torsten Sulima für die zahl- und ergebnisreichen Diskussionen zu verschiedensten Bereichen dieser Arbeit und die unschätzbare Unterstützung bei administrativen Vorgängen rund um Personalie, Bestellungen und IT.

Ulrich Abelein, Helmut Lochner und Dr. rer. nat. Stefan Simon, alle ehemals AUDI, für die enge projektnahe Betreuung des Themengebiets, die Einführung in die Zuverlässigkeitsarbeit in der Automobilindustrie und die hervorragende Gesellschaft auf Konferenzen.

Dr.-Ing. Maciej Wiatr, Mahesh Siddabathula, Dr. rer. nat. Dieter Lipp, Dr.-Ing. Gernot Krause und das Reliability Team bei GLOBALFOUNDRIES für die schöne und lehrreiche Zeit in Dresden, die Einblicke in die Halbleitertechnologiequalifizierung und die Zurverfügungstellung von Testwafern und Qualifikationsdaten, die diese Arbeit sehr bereichert haben.

Univ.-Prof. Dr. techn. Linus Maurer für das Angebot einzelne Messungen meiner ersten Veröffentlichung an seinem Messplatz durchführen zu können.

Silke Boche für Probenherstellung und die erfrischende Gesellschaft auf Fachmessebesuchen und in der Teeküche.

Peter Frank, Maximilian Reindl und das Werkstattteam für die allzeit schnelle, zweckmäßige und nachhaltige Hilfe bei den Erweiterungen des Messplatzes.

Dr. rer. nat. Ursula Goßner für das Defektätzen einzelner Proben und das wirkungsvolle Instituts-Desinfektionsmittel während der Coronakrise.

Eva Schober für die Erstellung der SEM-Bilder.

Isolde Belz für die freundliche Hilfe in allen verwaltungstechnischen Belangen.

Adnan Alsioufy und Thomas Lehndorff als meine Mitstreiter in der Zuverlässigkeitsforschung für die kollegiale Zusammenarbeit.

Simon Edler, Jonas Mair und Zhaohai Jiang für die außerordentlich angenehme Atmosphäre im Büro trotz Südlage.

Alle ehemaligen und aktuellen Doktoranden und Postdocs, sowie alle Mitarbeiter des Institutes für das gute Arbeitsklima und die schöne Zeit während der Doktorarbeit.

Die weiteren Kollegen und Partner der Infineon Technologies AG, der Robert Bosch GmbH, dem FZI Forschungszentrum Informatik und der HOOD GmbH im "autoSWIFT" Projekt für die erfolgreiche gemeinsame Projektarbeit.

Meine Familie, die mich bei allem unterstützt, was auch immer auf mich zukommen mag.

Meine Frau Stefanie für die unbegrenzte Ausdauer und Kraft, die du spendest. Du bereicherst mein Leben auf so vielfältige Weise!

www.ingramcontent.com/pod-product-compliance
Ingram Content Group UK Ltd.
Pitfield, Milton Keynes, MK11 3LW, UK
UKHW021653190726
13853UKWH00001B/227